Routledge Revivals

Housebuilding, Planning and Community Action

First published in 1986, *Housebuilding, Planning and Community Action* was written as an examination of the conflicts and tensions resulting from private sector housing growth in Central Berkshire, part of Britain's 'Silicon Valley' along the M4 motorway.

The book provides a detailed consideration of the various 'actors' and their interactions and explores the fight from Community groups and parish councils to halt development, in opposition to the government's reluctance to discourage economic growth. It focuses on four groups closely involved in the production, allocation, and consumption of new housing: speculative house-builders, local planning authorities, parish councils, and community/residents' groups. The motivations and actions of each group are examined, and the tensions between them are highlighted, set within the context of central government attitudes towards planning and private housebuilding.

Housebuilding, Planning and Community Action has lasting relevance for those interested in human geography, and the history of housebuilding and planning.

Housebuilding, Planning and Community Action

The Production and Negotiation of the Built Environment

By John R. Short, Stephen Fleming and Stephen J. G. Witt

Routledge
Taylor & Francis Group
ROUTLEDGE REVIVALS

First published in 1986
by Routledge & Kegan Paul

This edition first published in 2021 by Routledge
2 Park Square, Milton Park, Abingdon, Oxon, OX14 4RN
and by Routledge
605 Third Avenue, New York, NY 10017

Routledge is an imprint of the Taylor & Francis Group, an informa business

Publisher's Note
The publisher has gone to great lengths to ensure the quality of this reprint but points out that some imperfections in the original copies may be apparent.

Disclaimer
The publisher has made every effort to trace copyright holders and welcomes correspondence from those they have been unable to contact.

A Library of Congress record exists under LCCN: 85019387

ISBN 13: 978-0-367-77204-8 (hbk)
ISBN 13: 978-1-003-17022-8 (ebk)

Book DOI: 10.4324/9781003170228

Housebuilding, Planning and Community Action

The production and negotiation of the built environment

John R. Short,
Stephen Fleming
and
Stephen J. G. Witt

Routledge & Kegan Paul
London, Boston and Henley

First published in 1986
by Routledge & Kegan Paul plc

14 Leicester Square, London WC2H 7PH, England

9 Park Street, Boston, Mass. 02108, USA and

Broadway House, Newtown Road,
Henley on Thames, Oxon RG9 1EN, England

Phototypeset in Linotron Times, 10 on 11 pt
by Input Typesetting Ltd, London
and printed in Great Britain
by T. J. Press (Padstow) Ltd
Padstow, Cornwall

Library of Congress Cataloging in Publication Data

Short, John R.

Housebuilding, planning, and community action.
Bibliography: p.
Includes index.
1. Housing—England—Berkshire. 2. Housing policy—
England—Berkshire—Citizen participation. 3. Housing
policy—England—Berkshire. I. Fleming, Stephen
Charles. II. Witt, Stephen J. G. III. Title.
HD7334.B45S54 1986 363.5'09422'9 85–19387

British Library CIP Data also available

ISBN 0–7102–0723–9

Contents

Acknowledgments

1 Introduction 1

The main agents 1
The arena 2
The context 3
The contents 5

2 The circumstances 7

Local government 7
The land use planning system 13
Development pressure in Central Berkshire 24
Emerging tensions 35

3 The housebuilders 37

Stages and agents in residential development 38
The general context 40
The company context 41
Land assembly 44
The production of housing 60
Timber frame construction 71
The marketing and selling of new housing 78
Provision of the customer's capital 83
Capital provision 85
The external contacts of housebuilders 88
Links with the planning system 92
Emerging trends in housebuilding 112

4 The local planning authorities 115

The councillors of Central Berkshire 116

The system in operation 139
Summary: little power but all the blame 159

5 Parish councils 163

Parish organization 165
Parish politics 171
Parishes and planning 181

6 Residents' associations 203

Group formation 205
Internal resources 222
External contacts 230
Consequences 238

7 Stories and tensions 240

A parish initiative 240
Postscript 246
Burghfield Common 247
Newhurst Nursery 258
Emerging tensions . . . again but finally 265

Appendix 1 The housebuilders 276

Appendix 2 Planning applications for new dwellings in Central Berkshire 1974–81 285

Appendix 3 The local planning authorities 290

Appendix 4 The parish council survey 294

Appendix 5 Residents' associations 295

Bibliography 298

Index 303

Figures

Figure 1.1	Central Berkshire in regional context	3
Figure 2.1	Local government in Berkshire pre-1974	11
Figure 2.2	Local government in Berkshire post-1974	11
Figure 2.3	Central Berkshire (boundaries/local authorities, towns, communications)	25
Figure 2.4	Agent interactions	36
Figure 3.1	Operational radii of housebuilders in Central Berkshire	62
Figure 3.2	The price of dwellings constructed by different sizes of housebuilder	69
Figure 3.3	Major residential sites in Central Berkshire	75
Figure 3.4	Planning application success rate of housebuilders in Central Berkshire 1974–81	109
Figure 3.5	A typology of housebuilders' approaches to the planning system	110
Figure 4.1	Simplified outline of the development control process	141
Figure 5.1	Typology of parishes within Central Berkshire	164
Figure 5.2	Committee structure of parish councils in Central Berkshire	168
Figure 5.3	Parish politics	174
Figure 5.4	Parishes and planning policy	190
Figure 5.5	Parish council observations and district council decisions in Newbury (East) 1978–9	197
Figure 6.1	The cumulative national growth of amenity societies	204
Figure 6.2	Residents' groups in Central Berkshire	206
Figure 6.3	Residents' groups in Central Berkshire: the cumulative picture	207
Figure 6.4	Life cycle of residents' groups	220
Figure 6.5	Residents' groups per 1,000 population by parish	221
Figure 6.6	Residents' groups in Reading	223
Figure 6.7	Residents' groups in Wokingham	224
Figure 6.8	Local linkages of selected residents' groups	232

Tables

Table 2.1	The main DOE circulars involving builder-developers and land availability	20
Table 2.2	Floorspace costs in 1982: comparison of London and Reading	28
Table 2.3	Use of office blocks in Reading	28
Table 2.4	Summary of advantages and disadvantages of the alternative strategies	32
Table 2.5	Local groups' responses to initial structure plan proposals	34
Table 3.1	Stages and agents in residential development	39
Table 3.2	The sample of housebuilders	41
Table 3.3	Ownership of housebuilding companies	41
Table 3.4	Administrative staffing of a major housebuilding subsidiary company	44
Table 3.5	Sources of land information: number of firms claiming to receive information from various sources	50
Table 3.6	Sources of land purchase: number of firms claiming to use different sources	51
Table 3.7	Housebuilders' perception of the adequacy of the size of their land bank	56
Table 3.8	Operational radii of different sizes of housebuilder	61
Table 3.9	Entry to and withdrawal from Central Berkshire of the large companies according to the dates of their first residential planning applications	65
Table 3.10	Number of staff (excluding site labour) employed by housebuilding companies	68
Table 3.11	Neighbourhood names in Lower Earley	80
Table 3.12	Sample of incentives offered by housebuilders in Central Berkshire in 1982	81
Table 3.13	Companies with building society or bank mortgage quota agreements	83

Table 3.14	Sources of funding for housebuilding	86
Table 3.15	Possible points of contact between housebuilders and the planning system	93
Table 3.16	Number of housebuilders who have appealed by local inquiry	96
Table 4.1	Party affiliation in each of the districts of Central Berkshire and on Berkshire County Council at April 1983	116
Table 4.2	Age and political party of surveyed district councillors	119
Table 4.3	Interests of councillors and their spouses in local authorities of Central Berkshire as at May 1983	123
Table 4.4	Likely and unlikely combinations of motivations and roles: survey assessments in brackets	128
Table 4.5	Proportion of councillors attributing high and low levels of importance to various influences acting upon them	133
Table 4.6	Delegated items during three cycles of the four Central Berkshire planning authorities	142
Table 4.7	Breakdown of councillors' activities	149
Table 4.8	Proportion of applications (all and new dwellings) in each mode of committee response	152
Table 5.1	Responsibilities of parish councils under 1972 Act	166
Table 5.2	Contesting of parish council elections in Central Berkshire 1979–83	172
Table 5.3	Party politics and election contests in parishes of Central Berkshire	173
Table 5.4	Candidates and elected councillors in the parishes of Bracknell DC, May 1983	175
Table 5.5	Occupations of parish councillors	177
Table 5.6	Roles and motivations of parish councillors	177
Table 5.7	Parish council observations on planning applications	186
Table 6.1	Residents' group distribution by parish type	220
Table 6.2	Membership distribution of residents' groups	226
Table 6.3	Predominant occupational status of membership	227

Table 6.4	Dominant tenure type of membership	228
Table 6.5	Residents' groups and the planning system	233
Table 7.1	Estimates of costs of alternative levels of population growth (1975 prices)	267
Table 7.2	Planning applications and planning gain agreements in Wokingham District Council	268
Table A1.1	Classifying the sample	280
Table A1.2	Comparison of housebuilder size distributions: the interview sample, the Central Berkshire sampling frame and the national situation	281
Table A1.3	Classification of housebuilders by size and stage of dynamism	282
Table A2.1	New dwelling application decisions issued in last quarter of 1981 in Central Berkshire and in England	285
Table A2.2	Residential planning applications for Central Berkshire districts 1974–81	287
Table A2.3	Success rate (%) of new dwelling applications, by size and type, in Central Berkshire (excluding Reading Borough Council)	288
Table A2.4	Success rates of new dwelling applications in Central Berkshire 1974–81, by district and size	288
Table A4.1	The parish survey in relation to the parish development typology	294
Table A5.1	Breakdown of the resident group surveys	297

Blocks

Block 1	Office development in Reading	28
Block 2	The very small housebuilder	45
Block 3	The small housebuilder	53
Block 4	The medium-sized housebuilder I	64
Block 5	The medium-sized housebuilder II	73
Block 6	A large housebuilder	88
Block 7	How to get past the planners	100
Block 8	The registers of interest	123
Block 9	A policy administrator	128
Block 10	A rural Liberal councillor	130
Block 11	An urban Conservative councillor	131
Block 12	Pacifying ward pressure	136
Block 13	An example of applicant pressure	140
Block 14	District councillor activities and time commitments	149
Block 15	The composition of two parish councils	180
Block 16	A week in the life of a parish chairwoman	182
Block 17	Parish council observations and the district response	196
Block 18	A stopper group member	211
Block 19	A getter group member	213
Block 20	A stopper-getter group member	217
Block 21	Changing aims and roles of a residents' group	219
Block 22	The fight against development	235
Block 23	Community group activities and linkages	237
Block 24	'You win some, . . .'	239
Block 25	Planning gain 1 – Seeking community facilities	269
Block 26	Planning gain 2 – Tactics and pressure	270

Acknowledgments

Acknowledgment is firstly due to two other members of the project team, whose work is incorporated within the book. Catherine Starrett (PhD postgraduate) and Carey Roberts (Research Officer) jointly undertook the questionnaire survey of community groups in Central Berkshire and the tabulation of results. Ms Starrett subsequently analysed the data and produced a report which forms the basis of Chapter 6; Ms Roberts was involved with the questionnaire survey of councillors and the initial survey of planning committee meetings, and also drafted part of Chapter 2.

We wish to express our thanks to the many individuals and organizations who provided information and assistance during the course of our research, in particular the planning departments of Berkshire County Council, Newbury and Bracknell District Councils; Councillors Rosemary Sanders-Rose, Margaret Gimblett, Brain Fowles and Ron Williams; Gareth Capner of the Barton Willmore Planning Partnership; and Andrew Bennett, who was Land and Housing Officer for the Southern Region of the Housebuilders' Federation.

Within the Geography Department at Reading University, we are grateful to Chris Howitt and Sheila Dance for the production of the diagrams, and to Chris Holland, who had to type a number of drafts of the book.

1 Introduction

What we need now are not trans-historical theories of society but rather theorized histories of social phenomena.

(Castells, 1983)

Between January 1974 and June 1982 1,168,957 private sector dwellings were constructed in Britain. This was not a random process nor the simple result of pure market forces. Neither did it occur by official *diktat*. The amount of housing, its location, density, and design were the function of market forces and the product of negotiation between different interests. The process of negotiation is the subject of this book.

The main agents

There are a whole range of institutions and interests – let us call them agents – involved in the production of housing. Some are more important than others. National government, for example, in setting the economic climate and providing the planning ground rules is more important than a small residents' association. In this study we will concentrate on only a few of the agents. We will select those which allow us to consider the two main themes of state action – the need to facilitate capital accumulation while also ensuring legitimacy. In a capitalist democracy the state tries to meet the needs of capital while also maintaining the consent of the majority. In particular, the state is lobbied by builder-developers and by various community groups; it seeks to facilitate the operation of the producers of the built environment while being relatively open to the articulated voice of the consumers of the built environment. The conflicts between the different agents and the state's balancing act in meeting these often competing claims is a major theme of this book. The agents we have identified for analysis are those which encapsulate this accumulation/legitimation dichotomy. In particular, *housebuilders* who build dwellings, the *members and officers* of the local councils which give planning permission for housing construction and various *community groups* who represent local areas. These agents interact in a number of ways but the most important being within the procedures and institutions of land use planning and development control which we will lump together under the term the planning system. The process of negotiation is one of structured interaction, a bargaining within certain sets of changing rules. These interactions can be defined as a game in so far as games are rule-governed behaviours where outcomes are uncertain. By using the terms *game*, *rules* and *bargaining* we do not intend to imply that the negotiations at any one time are fair or even. Inequalities in resources and power exist and the rules aid some agents rather than others.

Moreover, as Gregory (1982, p. 16) points out:

> no game is ever a simple working out of deep-seated structural rules: the players' actions are reflexive – they are motivated and rationalised – and it is through the flow of move and counter-move that these rules are affirmed, infringed and challenged.

Each agent has different sets of interests, constraints and opportunities afforded by their position in the planning system and within the wider society. In order to understand this interaction it is necessary to look at the source of these constraints, the nature of the opportunities and the separate and shared ideologies which underpin action and behaviour.

The arena

The study is concerned with the interaction between agents in a particular part of the country – Central Berkshire, an area with a population of 373,300 in 1980. The choice is not arbitrary. We were based at the University of Reading and this was the planning area closest to home. In any study which uses one case study there is always the problem of the unique which makes it difficult to generalize. Our results are generated from Central Berkshire which in comparison with the rest of the country has seen high levels of economic growth and low levels of unemployment. Thus, it is not a typical area. But because of the strong development pressure it exhibits in heightened form the sorts of tensions less evident elsewhere. The area shows more clearly than most the differing interests involved in the planning system.

Too much human geography has been guilty of accepting spatial subdivisions especially at the sub-national level which relate more to data collection convenience than to theoretically informed class-

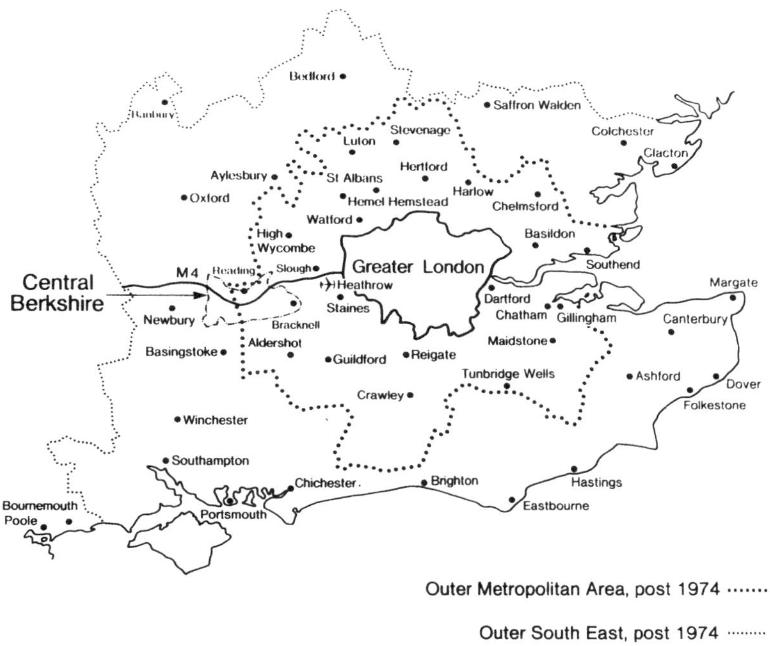

FIGURE 1.1 *Central Berkshire in regional context*

ifications. Central Berkshire is not a chaotic concept in this manner. It is a valid object of analysis because in being an area of structure plan preparation, it is an identifiable and real arena for the articulated interests of various agents.

The context

Our work is guided by three debates. The first is the general debate within social theory on the relationship between social structure and human agency. A number of writers, especially Giddens (1981), Bourdieu (1977, 1980) and Bhaskar (1979), have called attention to the failings of much existing social theory. They point in their different ways to:

(a) the *functionalist fallacy* which reads off actions and events from a preconceived structure, e.g. the functionalism of structural marxism which sees agents merely as bearers of predetermined positions assigned by a mode of production articulated in a social formation;

(b) the *dangers of voluntarism* which sees society as simply the product of human intentions.

There is now a common aim which seeks to provide social theories which can capture the relationship between social structure and human agency without degenerating into voluntarism or structural functionism. Gregory (1981) and Thrift (1983) discuss these issues with respect to human geography. We will not address this debate directly but our work is informed by the general need to consider the recursive relationship involved in the contingency of human agency unfolding within the choices, opportunities and constraints afforded by a transforming social structure. More specifically, we will seek to show:

(a) the evolving relationships between the main agents within the changing set of rules in the planning system;

(b) the ways in which the operational rules have themselves been transformed in the process of interaction.

In a very real sense we see the planning system as an institutional point of contact between agency and structure. The evolving recursive relationship between the planning system and the main agents is the central focus of this book.

The second debate is the continuation of the work of the senior author on the relationship between state, space, capital and community. These themes have been combined in different ways at various scales in two previous books. In *An Introduction to Political Geography* (Short, 1982a) the relationship between the changing world order and the nation state was considered while *The Urban Arena* (Short, 1984) took the analysis along a more specific route by examining the unfolding tensions in post-war Britain between capital and community in and through the state. This book carries on the debates by examining some of these tensions in a specific place at a specific time. By analysing agents and their interaction on this more detailed scale it is intended to flesh out debates in the previous books. This specific case study draws upon the ideas of the previous books but by looking at actual agents in real time it focuses attention on contingency, change, creativity and the relationship between beliefs and action.

Finally, we have been working within a well-established academic field. There is a small mountain of books and a vast array of papers on analysing different aspects of the British planning system. We have drawn on some of this literature but as outsiders coming into this field we were both overawed and disappointed; overawed by our lack of knowledge especially in the face of the amount of material – there are senior researchers in this field who have internalized more than we can articulate – but disappointed by the nature of much of the literature. As outsiders we looked in vain for many studies which allowed hypotheses about the effects of the British land use planning system to be sustained. Much of the planning literature ignores the need to demonstrate rather than assert the real consequences of the system. Breheny (1983, p. 114) blames

> the theorists who have been excessively abstract in their work, have ignored yet managed to misrepresent practice, and who have consciously failed to offer any presumptive advice to practitioners. The latter for their part have continued to lack any sustained critical assessment of their own activities and have lapsed into insular pragmatism. . . . The result has been theory which, albeit sophisticated, has become increasingly useless, and practice which has become increasingly devoid of intellectual credibility.

In this study we have adopted a naive realism which makes few assumptions and seeks to demonstrate rather than assert. While old hands may find our naivety at best innocent and at worst over-simplistic, newcomers will hopefully be enlightened and it is for them that the book is intended. The work is provisional. We present a case study at one point in time and basic empirical data on the main agents and the changing nature and forms of their interaction. We hope that other researchers will come along the same path but will go far beyond the efforts presented here.

The contents

The structure of the book is relatively straightforward. Since we are concerned with providing an outline of the system of interaction between the different agents, it is important to look at the agents in some detail. Chapter 3 thus considers the housebuilders, while Chapter 4 examines the functioning of local planning authorities (LPAs) in Central Berkshire paying particular attention to the position of elected members. Chapter 5 looks in detail at the role of parish councils in local planning affairs, organizations positioned between the formal institutionalism of LPAs and the informality of social movements. Chapter 6 considers the importance of residents' groups in planning matters. These four chapters thus consider the main agents concerned with the production, management and consumption of the one element of the built environment. Chapter 7 attempts to pull together these separate threads of the story in two ways. On the one hand three separate case studies will be discussed in depth showing how the interaction between the agents is full of 'accidents and conjunctures and curious juxtaposition of events' (Butterfield, 1950, p. 6). On the other hand the interaction will also be considered in a more analytical fashion by showing the emerging tensions implicit and explicit in their relationship.

In the exposition we will seek to provide particular examples while also drawing out generalities. We wish to generate theorized histories and and also theorized geographies. To those who require the steady pace of a general text or the rich diversity of only case studies then the exposition presented here may seem of uneven pace. We ask the reader's indulgence. The text should be seen as two elements of a whole, like beads on a string. The generalities run like a thread through the more discrete case studies. One gains stability from the one and substance from the other.

People, to paraphrase Marx, make their own histories and their own geographies. But not, he went on to add, in circumstances of their own making. In the next chapter we will consider some of the circumstances in the making of the recent history and the transformation of the human geography of Central Berkshire.

2 The circumstances

'I am the very slave of circumstance. And impulse.'
(Lord Byron)

The three main circumstances which provide the context for the unfolding relationships between the three main agents of developers, planning authorities and community groups in Central Berkshire are:

(1) the role and scope of local government;
(2) the land use planning system;
(3) the nature of development pressure in Central Berkshire.

Let us examine each in turn.

Local government

After April 1974 the elected representatives of the people of Berkshire had to operate a new local government system. In the rest of this book we will be concerned with the workings of this system. In this section we will examine the general background to this change, some of its causes and consequences and the more important subsequent events.

Local government under pressure

The 1972 Local Government Act, which signalled the reorganization of local governments in England, was the outcome of a whole series of pressures and attempts to change the system which had been operated relatively unchanged since 1888. In the old system the town-country split had a legislative form in the county borough/county council division which also marked a rough political divide between Conservative rural areas and Labour cities. A major source of conflict came from an attempt by some of the boroughs to expand their boundaries in order to meet land use pressures. The counties resisted this annexation which for them meant a loss of territory, population and rateable value. Despite the resistance 123,000 acres were annexed by county boroughs from counties between 1937 and 1958. The pressures increased throughout the post-1945 period, and a Local Government Commission was set up by the 1958 Local Government Act. The terms of reference of the commission set up by a Conservative government made sure that the county boroughs had to make very special cases for the extension of their boundaries. Moreover, the comission could only recommend; the final decision was up to the minister. Both Labour and Conservative ministers 'delayed, watered down and rejected recommendations made by the Commission' (Alexander, 1982, p. 18) in line with political expediency.

Although a messy business, boundary changes on their own did not necessarily call for root and branch reform. More important was a growing realization in the post-1945 period that the existing local government system was not performing well. The post-war years saw an enormous growth in state involvement at both central and local level. Since local government was called upon to shoulder some of the burden of spiralling state involvement many central government departments wanted local government units which could handle this new expanded role in an efficient manner. The Department of Education wanted education authorities with a minimum population of 500,000 while the Ministry of Housing and Local Government wanted city region type authorities for efficient planning. There was also dissatisfaction with the way local government was working. At the local level, the Association of Municipal Corporations wanted larger units and more power while at the central government level it was most clearly expressed in the setting up of the Maud and Mallaby committees in 1964 to examine local government staffing and the overall management of local authorities. As the Maud Committee noted, 'Parliament, Ministers and the Whitehall Departments have come increasingly to lose faith in the responsibility of locally elected bodies' (Maud, 1967, vol. I, para. 4).

By the 1960s then, central government was perceiving existing local government structures as inadequate: they were seemingly unable to meet the extra demands required by the growing role of the state. The exact reason for Richard Crossman setting up the Royal Commission on Local Government in 1966 probably owes as much to specific political contingencies as to a general attempt to restructure the lower levels of the British state system. But there was very little dissent from the decision and most critics attacked the form of reorganization, not the need for it. The old system had been discredited.

The reorganization of local government

The Royal Commission set up in 1966 was established by the Labour minister Richard Crossman under the chairmanship of Lord Redcliffe-Maud. Crossman had attempted to get a chairman committed to radical reform and sympathetic to the Labour party, but one possible candidate declined and Redcliffe-Maud was chosen after the prime minister had overruled other candidates. Like its chairman, the report was to be a compromise.

The work of the commission was based on three principles: the need to overcome the town/country division; the need to create local government units which could achieve economies of service provision (in practice this meant units of between 250,000 and 1 million); and the need to avoid where possible the division of responsibility between different local government levels. The solution proposed in the majority report was for eight regional councils, three metropolitan authorities with a two-tier structure, and fifty-eight unitary authorities responsible for most local government services (Redcliffe-Maud, 1969, vol. 1).

Reaction to the proposals varied. There was not much support for the regional councils and these were not to appear again on anyone's agenda. The Labour party supported the other main proposals and in their White Paper the following year they kept the unitary structure suggesting only the addition of another two metropolitan areas. After the 1970 general election it was the Conservatives who implemented re-organization. The Tories' White Paper published in February 1971 went against the unitary principle in seeking to retain a two-tier system of county and district with six metropolitan areas. Behind this White Paper lay the operation of local lobbying and the entrenched power of Conservative rural interests. As many traditional counties were retained as possible and greater power was given to district councils.

Seen in terms of the aims and principles of the Redcliffe-Maud Commission the final result of local government reorganization was disappointing. The unitary authority principle was sacrificed to local interests and the result was to strengthen the rural areas. Tight boundaries were drawn around metropolitan areas, and the large towns which were not made metropolitan areas (e.g. Bristol) saw a loss of independence with incorporation into the counties. Within counties, rural and municipal districts gained legislative parity with former county boundaries. Implemented under a

Conservative government local government reorganization retained the Conservative rural power base. Labour votes were channelled into metropolitan areas and in the counties the power of large towns was diluted. Commenting on the outcome Johnston (1979, p. 150) notes, 'the final map bears the imprint of political design, and in general it has worked to Conservative advantage'. Local interests prevailed over the larger city regions or unitary authority solutions.

The change in Berkshire is shown in Figures 2.1 and 2.2. In the pre-1972 period Berkshire consisted of the one county borough of Reading, six municipal boroughs, eleven rural districts and one urban district. With reorganization there was a significant loss of territory to the South Oxfordshire District but the elongated nature of the county remained, a result of historical precedent and rural lobbying, its cumbersome nature expressed in the division of the county into three structure plan areas of East, Central and West. In this book we will be concentrating on Central Berkshire which consists of Bracknell, Wokingham, Reading and Newbury East. In the new six-district system tight boundaries were drawn around Reading and Slough where the local Labour parties did not want the boundaries extended to incorporate the non-Labour-voting suburbanites, while the adjacent communities in turn did not want to be incorporated into the higher-rated towns. However, given the reality of the Reading urban system with large journey-to-work and journey-to-shop flows across the boundaries and severe land use pressures within Reading's boundaries, this division has remained problematic. In 1983 local public opinion was again aroused in the adjacent areas in opposition to an attempt by Reading to extend its boundaries, an attempt led by the Chief Executive and most strongly supported by Conservative and Liberal members of Reading Borough Council.

Local government unreformed

The 1972 Act transformed most levels of local government in England. There was one level however which remained relatively untouched, the parish. It has a long history in England. Initially an early church administration unit, parishes began to obtain non-ecclesiastical functions as early as 1600. But it was the 1888 and 1894 Local Government Acts which established parish councils as administrative units. In areas with more than 300 inhabitants they were to be responsible for burial grounds and public parks. In its review of local government, the Redcliffe-Maud Commission noted the value of parishes as a vital element to democratic local government. Indeed in their plans for boundary reorganization

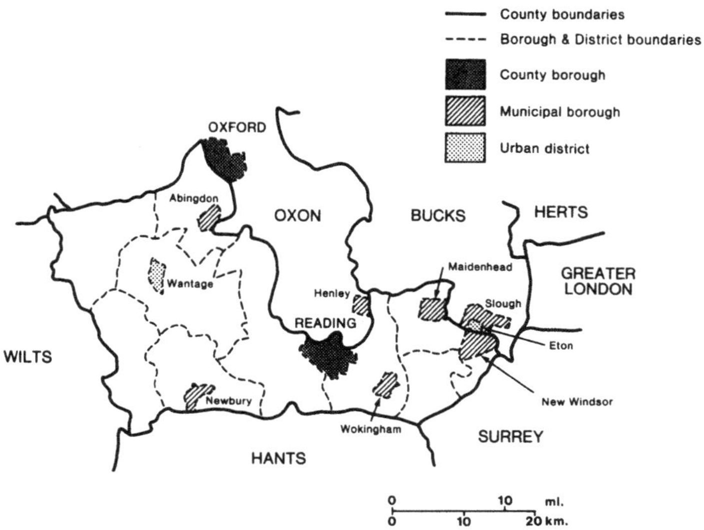

FIGURE 2.1 *Local government in Berkshire pre-1974*

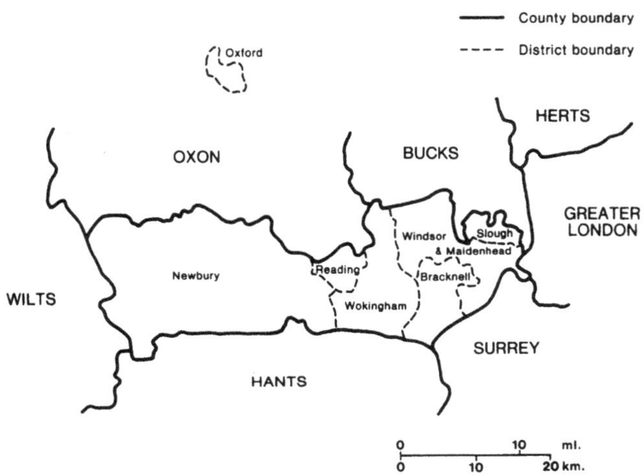

FIGURE 2.2 *Local government in Berkshire post-1974*

they noted that in no circumstances would they divide the parish. The Commission's recommendations were incorporated in the 1972 Act which strengthened parishes giving them the power of discretionary expenditure (under section 137) up to 2p in the pound rate on anything that could benefit the community and the right to be notified of planning applications in their parish. For central government, the parishes contained all the elements of an ideal local government system: combining the ideology of democracy, participation and conflict-free localism with the reality of limited expenditure and political impotence.

Central/local relations in hard times

The relationship between central government and local authorities became increasingly tense throughout the 1970s. In essence the friction arose from two related aspects. The first and most important was the attempt by central government to reduce local government expenditure. In the 1970s, local authorities were responsible for just over a quarter of all government spending. Local authorities were primarily responsible for such welfare functions as education and housing. As Britain's economic crisis increased central government sought to control and redirect such expenditure. It was not an accountant's exercise but an exercise in political control and spending prioritization. Labour local authorities in particular were more responsive to local pressures and demands for welfare type spending while the prevailing ideology in Whitehall began to see such expenditure as wasteful.

The ability of central government to control local government came through the funding of local authority expenditure. Local spending grew in the 1960s and 1970s but there were limits to the financing of this expenditure from the local tax base of property rates. By the mid 1970s, central government was providing, in the form of grants, over a half of all local authority spending. From 1975 onwards, cash limits were introduced and local authorities were no longer given supplements to their grants. Under the Conservative government of 1979 with its commitment to tight financial controls and a massive redirection of public spending the central/local relationship was put on a radically different footing. The 1980 Local Government, Planning and Land Act placed controls on capital spending and put the fixing of the grant to local authorities in the hands of the Secretary of State. In the 1981 Local Authority Act, greater powers were given to the Secretary of State to penalize local authorities deemed to have overspent. Both Acts signalled a dramatic shift towards greater centralized control.

The second facet of the central/local relationship has been in attitudes towards economic development. In the 1960s economic growth seemed assured and the question of its distribution seemed more important. Labour governments in particular pursued strong regional policies since much of their political support came from peripheral regions. But, as the recession deepened in the 1970s, economic growth itself could not be taken for granted. Questions of distribution declined in importance as emphasis was placed on stimulating economic growth. On the political left this was seen in terms of protecting industry and on the right in terms of reducing government controls. A Conservative government elected in 1979 weakened regional policy measures and introduced enterprise zones as part of an overall approach which saw the state itself as part of the problem. The switch was marked in the land use planning system as a series of government circulars exhorted local authorities to be more receptive to the needs of the private sector. As we shall see later in this book, central government has backed up its words with action in overturning local authority planning proposals and development control decisions.

In effect the late 1970s marked a profound shift in central/local relations. Power became more centralized and this power was being used to reduce expenditure and to facilitate market forces.

The land use planning system

Local authorities have many duties and functions to perform. The one which we will be examining in some detail in subsequent chapters is their planning function. In this section we will make some general observations; later chapters will examine the particulars of land use planning in Central Berkshire.

Land use planning, by which we mean the set of rules and procedures arising from the appropriate state legislation, is more than just a bureaucratic exercise. It is at one and the same time an ideological discourse on the nature of the built environment and a system of negotiation which sets the rules of access and institutional purchase afforded to different agents. It is ideological in the sense that ideas about planning emanate from specific agents, albeit mediated in the process of articulation; and these ideas both interpret and justify social reality. Builder-developers, for example, argue that an 'efficient' planning system can only be achieved by increasing land release, something which also aids their own specific operation. These ideas can be sincerely held and the use of the term ideology does not imply that they are deliberately falsified. Planning also involved the cooperation of and controls on sets of agents concerned with the built environment. The precise formulation of planning sets the degree and type of cooperation and coercion afforded to different agents. The agents in turn seek to improve their access. Builder-developers, to use them again as an example, have sought to get their views on land availability incorporated directly into planning procedures at both the national and local level. It is in these senses that we can see planning as an ideological discourse and a system of negotiation.

The land use planning system which currently operates in contemporary Britain is a product of incremental changes over the years. However, we can identify an ideological rupture between the pre-1940 and post-1940 periods. We will thus concentrate on the evolution of planning in Britain in the post-1940 period.

Planning for reconstruction

In 1940 the Cabinet set up a War Aims Committee. It was challenged with making suggestions about the post-war European and world system and 'to consider means of perpetuating the national unity achieved in this country during the war through a social and economic structure designed to secure equality of opportunity and service among all classes of the community' (quoted in Cullingworth, 1975, p. 4). The reconstruction machinery set up by the committee was part response to Nazi propaganda and part planning for reconstructing a war-damaged Britain, along lines which incorporated some measures of radical change but which also maintained the legitimacy of the existing system. The physical reconstruction of the built environment figured large in these deliberations. And in 1941 a recommendation to the Cabinet proposed the setting up of a central planning authority, the control of building development and the public acquisition of land with limits to compensation. Political debates at the time were informed by various reports. The war years witnessed a remarkable flurry of government activity as a whole series of reports were produced – the Scott Report on Land Utilization in Rural Areas (1942), the Uthwatt Report (1942) on Compensation and Betterment, the Reith Report on New Towns (1946) and the Dower Report on National Parks (1945). It was the Uthwatt Committee which gave the main lead in debates about land use planning. This committee, set up because of a recommendation in the influential Barlow Report on the Distribution of the Industrial Population published in 1940, called for the nationalization of much of the country's land and the payment of a regular betterment levy by landowners. In effect it was a proposal for land nationalization.

The post-war land use planning system was officially created with the passing of the Town and Country Planning Act of 1947. It was one of a series of Acts – the others included the Location of Industry Act (1945), the New Towns Act (1946), the National Parks Conservation and Access to the Countryside Act (1949) – which gave substance to many of the wartime reports. The recommendations were turned into legislation by the first post-war Labour government given a huge mandate for widescale social reform. Under the 1947 Act development rights were nationalized, and private owners of land had to seek government approval (planning permission) to develop their land. No compensation was payable to landowners who were refused planning permission and a betterment levy of 100 per cent on the development gain of land given planning permission was to be charged by the Central Land Board. The counties and county boroughs were charged with operating the system. A development plan was to be prepared by each authority which was to guide development within their area and serve as a basis for development control, i.e. the granting or refusing of planning permission. The final say, however, lay with the central government minister responsible for planning since he had the power to amend plans and overturn development control decisions on appeal. This then was the planning machinery established to reconstruct a Britain emerging from the rigours of war. It was a machinery marked by compromise. The initial Uthwatt Report had effectively called for land nationalization but the planning proposals came up against the ideological sanctity of private property and the power of the property lobby. Between the first Bill in 1947 and the Act there had been a series of official interdepartmental reports, wrangling in Cabinet and parliamentary revisions, most of it centring on compensation to owners. As one Treasury official of the times noted,

> The State leaves the actual ownership of the land in private hands but takes away the profit motive, the mainspring of private enterprise, by nationalizing the development value. If I were labelling this scheme, I would call it 'bastard Tory Reform'. . . . We feel that the scheme should go a little further in the way of socialization or should not go so far as to take the profit incentive out of private ownership. Any in-between system is likely to get the worst of both worlds. (quoted in Cullingworth, 1975, p. 259)

The reconstruction of planning

The 1947 Act provides a basic reference point. It was the template for subsequent changes in post-war planning systems. These changes were many but we can identify three significant ones.

(i) PUBLIC PLANNING AND PRIVATE GAIN

As the Uthwatt Committee had noted, in a system of private ownership, landowners could benefit from planning decisions. This has been termed betterment or windfall. In other words there was private gain from public planning. Attitudes to this gain varied according to political persuasion. On the left and in a long Liberal radical tradition there is the belief in not only public control over land but public gain from public planning. On the political right, by contrast, the emphasis has been on strengthening and maintaining property rights. These differences have been translated in planning legislation as changes in government have occurred. The compromise solution adopted by the 1947 Act was not to last long. The Conservative party came to power in 1951 and set about re-establishing the priority of private property interests in the planning system. The betterment levy was dropped in 1953, the role of the Central Land Board was much reduced and in 1959 compensation was paid to landowners on the higher basis of full market value. There then followed a sequence of attempts by Labour governments to control private gains from public planning and successive Tory amendments. In its 1967 Land Commission Act the Labour government charged a betterment levy of 40 per cent on development gain and established a Land Commission whose job it was to secure land cheaply and then release it for strategic planning purposes. The aim was to ensure that planning was not hampered by lack of cheap land. The Tory government of 1970 suspended the Land Commission a year after coming into office and repealed the betterment levy.

The next phase began with the 1974 Labour government which was committed to community gain from planning. Between the intentions of their 1974 White Paper and the reality of the 1975 Community Land Act fell the shadow of the powerful property lobby. The 1975 Act was hamstrung from the beginning by lack of finance, legislative power and political commitment at senior levels of government. The Tories discontinued this scheme on coming to power in 1979 and in their 1980 Local Government, Planning and Land Act directed local authorities to release more land for private development.

(ii) CHANGING CONTEXT AND NEW PROCEDURES

The 1947 system was established for a sluggish economy with little development pressure especially from the private sector. The 1950s and particularly the 1960s however were periods of economic growth and heavy development pressure. The system was becoming overloaded and the development industry was getting exasperated by the slowness of the system. By July 1951, which had been set somewhat optimistically as a deadline for submission of plans in the 1947 Act, only twenty-two of the 145 authorities had kept to the official timetable. It was not until 1961 that the entire country was covered by approved plans. A similar tale of delays applied to the supposedly quinquennial review of the plans and by the end of the 1960s when the development plan system was changed, there was considerable variation in the stage which various authorities had reached with their plans and in the extent to which they were operating non-statutory plans.

In May 1964 the Planning Advisory Group (PAG) was established by the Ministry of Housing and Local Government to undertake a review of the planning system. The report of the group concentrated on the development plans which were regarded as the main area for concern (PAG Report, 1965). PAG proposed that the single 1947 style development plan should be replaced by a new two-plan system. A structure plan would indicate the overall land use structure and policy intentions, and more detailed local plans for specific areas would promote and guide development at the local scale. Only the structure plan, containing the overall planning policy for the area, would be examined and approved by the minister. Once this had been done, local authorities could draw up and adopt local plans without reference to the minister, providing they conformed with the principles of the structure plan.

The 1968 Town and Country Planning Act enacted a number of the suggestions set out in the PAG Report. Under this Act development plans were to consist of the structure and local plan. The administration of the appeals machinery was also altered by the delegation of responsibility for deciding many of the minor cases to inspectors. It was hoped that 60–70 per cent of appeals would then be decided by the inspectors thus relieving much of the pressure upon central government and speeding up the whole system. Another change, aimed at quickening the pace of the planning system, introduced the discretionary power of LPAs to delegate to their officers the responsibility for some of their development control work. In subsequent chapters we will note the degree of delegation in the LPAs of Central Berkshire.

The 1968 Act was seriously affected by local government reform. The result of the 1972 Local Government Act was a two-tier system with planning functions divided between the two levels. Counties were to be responsible for broad planning strategy and districts for local planning and most development control matters. The rationale behind the new development plans of 1968 was thus removed since it could no longer be assumed that local plans would conform generally to the structure plan. Often the districts had very different views to the county planning authority as to how structure plan policies should be elaborated in their areas. The planning system which has operated since 1974 is therefore one in which there is great scope for friction between the two levels of local government especially over responsibilities concerning the new style development plans.

(iii) PARTICIPATION AND NEW RULES OF NEGOTIATION

The 1947 system has been transformed as a system of negotiation. Three aspects can be identified.

Public Participation The need for public participation in planning was first given official notice in the PAG Report. Set up to improve a system under pressure, the PAG Report introduced a notion of public participation, not so much as a participatory exercise but as a method for winning public support. By the early 1960s the smooth operation of the planning system was being 'held up' by a large number of appeals. Cullingworth (1981) describes a central government machinery crippled by the work caused by development plan approval and planning appeals. Far better, the PAG argued, to engineer public consensus at the early stages of plan preparation rather than be involved in long and costly amendments and planning appeals. Public participation would thus provide local authorities with a new opportunity for winning public support for their proposals.

The PAG Report was incorporated into the 1968 Town and Country Planning Act. It was acted upon because of the increasing level of public involvement in planning through community action and objections to planning proposals and as a response to the minor legitimation crisis faced by the government in the 1960s when there was an official recognition of public disillusion with government. The government response can be seen in a number of committees, e.g. the Redcliffe-Maud Committee on local government reform, the Ingleby Report on family advice centres, the establishment of the Home Office community development projects in 1969, all aimed at overcoming the lack of public credibility in governmental services. The development of formal public participation can be seen as a response to community action and as part of the official response to a crisis of credibility.

Prior to 1968 there was no statutory obligation for local planning authorities to involve the public in plan preparation. In the 1968 legislation, public participation was seen as a method of improving the planning system through reducing public opposition and incorporating public involvement by raising public consent for planning. As we will show later these aims are often contradictory. The 1968 Town and Country Planning Act conferred no new rights on citizens. Rather it put a statutory duty on local planning authorities to publicize their plan preparation and ask for public comments. A committee was set up under the chairmanship of Arthur Skeffington to consider 'the best methods of securing the participation of the public at the formative stage in the making of development plans' (Skeffington, 1969, p. 1). The Skeffington Report of 1969 is basically a handbook to local planning authorities of how best to publicize the proposals in the way that informs people and allows some public input.

The 1971 Town and Country Planning Act incorporated the spirit and some of the main recommendations of the Skeffington Report. This Act required local planning authorities to take steps to ensure that the public is made aware of their rights and opportunities to make representation. However the only statutory obligations relate to plan making. For the majority of people plan preparation is a rather abstract notion of little immediate concern. There is no statutory requirement for public participation in development control matters, a subject outside the terms of reference of the Skeffington Report. Yet the making of structure plans is often seen as the aspect of planning which is of the least importance and of least interest to the local community. They concern themselves more directly with development control issues.

Although the 1971 Act gave only restricted access and was originally seen as a method of incorporating protest, public participation has given some, albeit limited, institutional purchase to organized pressure groups. This institutional space has been utilized by the more organized pressure groups often to their advantage.

Builder-developers' participation There is a large amount of literature on the growth of public participation in the British planning system. There is a great deal less said or written about the participation of the builder-developers. The major reason lies in the different ways the rules of access for the various agents have evolved. In the case of public participation schemes have been introduced in legislative measures. Participation by builder-developers is less publicized and occurs not so much in the legislation but in the government circulars. These are issued at irregular intervals by the Department of Environment to LPAs. Technically, they are of an advisory nature only but they are powerful devices since failure to implement development control advice may lead to appeals. Decisions not in accordance with the circulars will in all likelihood be lost on appeal. Circulars are thus used to coordinate planning across the country.

A reading of the government circulars shows a steady increase in the way builder-developers' demands have been incorporated into the language and practice of land use planning. Table 2.1, for example, shows the relevant circulars associated with land availability. This has been a subject dear to the hearts of builder-developers as they have sought to orientate the planning system to release more land for development in order to ensure 'adequate' supplies of land for development. Such lobbying has been most successful with the incoming Conservative governments. There are many strong ties between the Conservative party and the construction sector. But, as Table 2.1 shows, land release has also been aided by Labour administrations.

TABLE 2.1: The main DOE circulars involving builder-developers and land availability

Circular number	
10/70	Land availability for housing, 14 December 1970 'The availability of land for development is essential to the stability of house prices and to the revival of the housebuilding programme'. (1) Release of land 'should be sufficient to meet housing needs . . . for at least the next five years' 'make generous additional releases . . . without detriment to other important planning objectives e.g. green belts' Phasing encouraged (2) Monitoring of the balance between housebuilding rates and planning permissions granted. Already undertaken in SE Region. (3) Acquisition and disposal of land by local authorities – LAs to dispose of surplus land for private housebuilding where this will not prejudice LA housing (4) Consultations with the building industry encouraged by initiation or response (5) Resources of DOE regional offices available for advice or assistance to aid above

102/72	Land availability for housing 17 October 1972 'appreciates the helpful response [to Circ. 10/70]. . . and the part . . . played in the local conferences with builders and landowners . . . [but] . . . not enough land is actually available to housebuilders in certain areas, mainly in the South-East.' (1) Five years' land supply again requested but defined: five times the average of the last five years or last year, whichever is greatest (i.e. past building rates) (2) Importance of safeguarding environment, esp. green belts, AONBs, and good agricultural land (3) Need for interim plans for some areas (4) Infrastructure and highways costs should not provide excuse – use of Section 52s encouraged (5) Time limits on planning permissions to be considered (6) Mapping of developable sites (three acres or more) (7) Land assembly by local authorities encouraged in SE (8) Loan sanctions to aid assembly (9) Compulsory purchase may be supported
22/73	Land availability for housing 1 October 1973 'supplementing and amplifying the advice in Circ. 102/72' (1) Monitoring and updating of 102/72 mapping (2) Local authorities asked to review further and dispose of land surplus to requirements (3) Loan sanction and compulsory purchase further encouraged (4) Guidelines for development control: Strong general presumption in favour of housing in growth areas Higher but not high densities encouraged Infrastructure not to be sole reason for refusal Need: need not always be demonstrated Design and layout should not cause such delay – planning brief encouraged
24/75	Housing: needs and action 25 March 1975 'a response to the urgent need to provide more housing for small households, and to speed up the provision of houses generally.' Considerable attention given to LA housing (1) Planning delay – priority given to housing planning application – DOE giving lead re appeals (2) LAs should allow higher densities
44/78	Private sector land: requirements and supply 15 June 1978 'Local Authorities have a permanent role in providing development land and in seeing that a sufficient supply . . . is made available for private housebuilding.' 'Local planning authorities think in terms of need: builders think in terms of demand. These two points of view have to be reconciled.'

	(1) Discussions between LAs and builders 'to get at the facts of land supply and the pattern of demand for housing in the area' Is the land: sufficient, serviced, and available for builders to buy? Will the houses be: marketable, and over what period of time? Emphasis: consultations assessing supply marketing constraints phasing and large sites districts to encourage diversification of builders
9/80	Land for private housebuilding 15 April 1980 'most concerned that the availability of land should not be a constraint on the ability of housebuilders to meet the demand for home ownership' (1) Active co-operation of planners and builders (2) Five-year figure now defined using structure plan policies 'ensuring that planning considerations are taken into account' (3) land should be 'genuinely available' (4) marketing considerations should be taken into account (5) Greater Manchester type studies encouraged
22/80	Development control – policy and practice 28 November 1980 '[The planning] system has a price, and when it works slowly or badly, the price can be very high and out of all proportion to the benefits.' 'Streamlining the planning system' (1) Speeding up of the system 'to play a helpful part in rebuilding the economy.' Eight weeks period to be more firmly adhered to (2) Development only to be prevented or restricted when this serves a clear planning purpose and the economic effects have been taken into account' Enforcement Aesthetic control reduced

The Conservative government voted in in 1979 has been particularly pro-developer. For example, the role of the Housebuilders' Federation (HBF) as a pressure group was positively encouraged by the then Secretary of State for the Environment, Michael Heseltine:

> The HBF has a key role to play. . . . I do happen to believe that your industry is now becoming organized effectively to do many of the things that I wish that industry at large had

> organized itself to do twenty years ago. . . . You are now showing a willingness to get involved in the real nuts and bolts dialogue with local authorities and central government. . . . I don't say you will get results immediately . . . but democracy is about pressure. People like me are a focal point of pressure. Indeed what politicians do in many ways is to oil the wheels, to arbitrate between the conflicting pressures, to take decisions about what democracy actually wants and will stand. (*Housebuilders' Federation Annual*, 1981, pp. 21–2)

Individuals such as Heseltine's personal advisor, Tom Baron, managing director of Whelmer/Christian Salvenson, also played an important part in forwarding the views of builder-developers (Rydin, 1983) and the achievements of the pressure and persuasion were to be seen in Circulars 9/80 and 22/80.

Circular 9/80 continued to pursue the land availability issue and called for the active cooperation of planners and builders while Circular 22/80 clearly marked a new phase: 'the Government wants to make sure that the planning system is as positive and helpful as it can be to investment of industry and commerce and to the development industry' (DOE, Circular 22/80, *Development Control – Policy and Practice*, 1980).

In effect then the planning system has been altered to reflect more closely the views and wishes of builder-developers. Another agent has been encouraged and incorporated more directly into the system of negotiation.

Local Planning Authorities and Planning Gain The architects of the 1947 system had not envisaged a practice of planning in which there would be formal bargaining between developers seeking planning permission and local planning authorities. The authorities' job was simply to grant or refuse planning permission. Recent years, however, have seen more explicit bargaining between the authorities and developers, and especially with the large developers. Under the original 1947 Act any agreements between local authorities and developers required ministerial approval and appeared only in the form of restrictive covenants. In other words local authorities could only tell developers what they could not do. Subsequent years have seen a change in the rules. The need for ministerial approval disappeared in the 1968 Act and under Section 52 of the 1971 Act local authorities were empowered to enforce positive covenants. In other words they could ask developers to do certain things, to provide certain facilities. The power of the Section 52 agreements was further strengthened by Section 126 of the 1974 Housing Act which ensures that covenants in

Section 52 agreements are enforceable against successors in title of the original landowner. Since then some authorities have given planning permission subject to conditions and agreements. In effect authorities are using their power to lever concessions from developers. Typical, for example, is a local plan which states that large new housing schemes should take into account play areas for children and/or provide a mix of housing types. In effect the LPA is attempting to obtain planning gain. For the local authority Section 52 agreements allow a securing of some public advantage from the granting of planning permission. For the developer such conditions may prove unacceptable if the financial viability of the project is put into question but on the other hand agreements may be the price to pay for planning permission which otherwise would be refused. The precise deal struck depends on the bargaining strength of the local planning authority and the developer.

To summarize: Within the framework of structure plan generation at county level and local plan generation and development control at district level, spaces for negotiation have been afforded to builder-developers in the practice of planning and to the 'public' at the stage of plan preparation. This is the system which operates in contemporary Britain.

Development pressure in Central Berkshire

Local planning authorities are more reactive than active bodies. They respond to planning applications, they rarely initiate them. The work of LPAs thus depends on the nature of local development pressure.

Development pressure

Central Berkshire, consisting of the districts of Bracknell, Wokingham, Reading and Newbury East (see Figure 2.3), is an area which has come under considerable pressure for much of this century. On the western fringes of the London region, Central Berkshire has attracted growth from the metropolis as well as generating indigenous development pressure.

Central Berkshire is part of that corridor of growth stretching along the M4 from West London to Bristol which has variously been described as Britain's Sun-rise Strip, the Western Corridor, and Silicon Valley, and has achieved the accolade of being the subject of academic articles (Breheny *et al.*, 1983), journalistic endeavours in *The Times* (24, 25, 26 January 1983), *The Econ-*

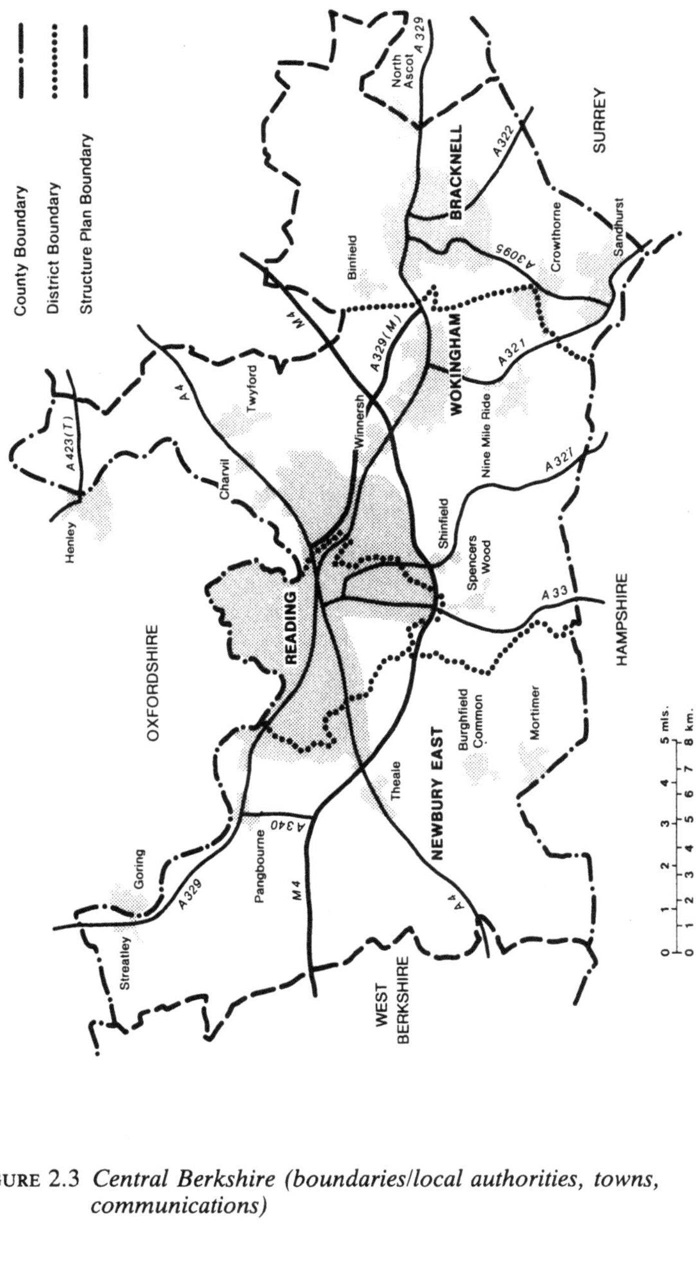

FIGURE 2.3 *Central Berkshire (boundaries/local authorities, towns, communications)*

omist (30 January 1982) and the *Guardian* (26 April, 1982), as well as the subject of radio and TV programmes.

Three types of development pressure can be identified in the area. First, there is the growing pressure for office space. On the demand side there has been a growing number of companies wishing to locate or relocate their offices in the main towns of Central Berkshire particularly Reading, Bracknell and to a lesser extent Wokingham. Because of the M4 motorway, high speed train services and ready access to Heathrow Airport these towns have good links with London, Europe and the rest of the world. There has been a significant number of moves by firms from London to Reading and Bracknell in search of lower overheads. Such international companies as Metal Box and Foster Wheeler now have their headquarters in Reading. Although rent levels in Reading are comparable with those in Holborn and Covent Garden rate bills are 60 per cent less. Firms wishing to relocate meet less resistance from the workforce than if they wanted to move to the Midlands, the North or Scotland, as Berkshire has an image of a rural attractive county. Given this sturdy increase in demand institutional investors and developers have been attracted by the development potential of urban sites in Central Berkshire. For the large investors particularly pension funds and insurance companies who have problems of how to invest millions of pounds, investment in property takes up large sums of capital in relatively profitable investments (Plender, 1982). The net result has been a flow of planning applications for office development in Central Berkshire, a flow which had by the late 1970s turned into a flood. The crude figures of planning permission fail to capture the reality of the process. In the case of Reading the centre of this quiet market town was completely transformed by the boom in office construction as many of the central city streets took on the feeling and sounds of a permanent building site. Many of the offices were occupied, and by the early 1980s Reading was the third largest centre for insurance after the City of London and Croydon, but vacancy rates by the third quarter of 1982 were 22 per cent. This relatively high rate was partly the function of the recession and partly a function of the conflict between the owners and users of office space. The owners wanted long-term leases while the users preferred shorter-term leasing arrangements with break clauses. Vacancy rates were concentrated in particular categories of offices. While the big companies wanted the large prestige sites and the smaller companies wanted the small offices, particularly conversions of historic houses, owners of office blocks which fell into neither of these categories experienced the greatest problems in getting tenants.

The second main pressure of development is in the form of demand for industrial premises, both manufacturing and warehousing. High-technology industries have been attracted and have grown within Central Berkshire. Two types can be identified. First, there are the computer and micro-chip firms such as ICL, Racal, Digital, Hewlett Packard and Micro Consultants all of whom have opened premises in Berkshire. Digital's European Research and Development section for example is now housed in purpose-built accommodation just outside Reading. Close to London and Heathrow, with a reputation of a willing workforce and a pleasant environment, Central Berkshire has become Britain's equivalent to Silicon Valley. This reputation has attracted more firms as a benign cycle of growth in high-tech industries has spiralled upwards. Second, there are the high-technology defence industries associated with weapons and defence system manufacturing. Often this division is blurred. Racal for example is both in computer technology and weapons research. Racal has been in Bracknell since the 1950s and of its 15,000 employees 5,000 now work in Central Berkshire. These war/defence-related industries have both grown *in situ* and been attracted for the same infrastructure and labour-related reasons as the more benign computer firms. There is also the nearby location of Defence Research Establishments. Situated around the pleasant Berkshire countryside are the Weapons Research Establishment at Aldermaston and the Royal Ordinance Factory at Burghfield, while in adjacent counties can be found the Atomic Energy Research Station at Harwell, the Nuclear Fusion Research Centre at Culham, and the Aircraft Research Establishment at Farnborough. If the swarming of computer firms has allowed commentators to label this area Silicon Valley, then the concentration of nuclear weapons establishments has allowed the peace movement to dub it the Valley of Death.

As with offices, the demand for industrial premises attracted the large investors and property developers. The BP pension fund for example has developed the Forbury Industrial Park in Reading, while the Legal and General Insurance Society funded a ninety-acre greenfield industrial/warehousing development in the Winnersh Triangle. In the corporate mind, Central Berkshire is seen as ripe for development.

In terms of industrial development this is an area of relative growth. Restructuring has taken place, as elsewhere in the UK, in which 'old' manufacturing jobs have been lost. The traditional manufacturing industries of this region, beer, biscuits and bulbs, halved their employment figures between 1971 and 1981 and the decline continues apace. In contrast to the rest of the county,

Block 1 Office development in Reading

Supply and demand
Recent years have witnessed an office boom in Reading. Office floorspace has grown from 1.3 million sq. feet in 1970 to 4 million by 1983. The offices have been constructed by speculative developers attracted to the perceived economic prospects of Reading and the Western Corridor which it is agreed answer a steady demand for office space. Most of the finance for these schemes comes from pension funds. The demand for office accommodation falls into three categories:

(a) the birth of new firms
(b) generated by established companies
(c) decentralizing/inmigrating firms from outside

Since 1970 only 30 per cent of the take-up of development space has been the result of decentralization. However, those firms which have migrated to Reading have shown a marked tendency to expand. The reason for movement to Reading, especially from London, are clearly illustrated in Table 2.2.

TABLE 2.2: Floorspace costs (£ per sq. ft) in 1982: Comparison of London and Reading

	City of London	*West London (e.g. Mayfair)*	*Reading*
rent	31.0	25.0	12.75
rates	15.0	9.0	2.50
services	7.0	4.0	2.0
total	53.0	38.0	17.25

A company requiring a 50,000 sq. foot office block could, therefore, holding everything else constant, save £750,000 per annum by moving from West London to Reading and £1,707,500 per annum by moving to Reading from the City of London.

The current distribution of users is as shown in Table 2.3.

TABLE 2.3: Use of office blocks in Reading

% occupied		*examples*
public sector	38	DHSS
industrial sector	25	Digital
financial sector	25	Coopers Lybrand
business and professional services	12	Hogg Robinson

Periodically the office market is oversupplied. Developers make their individual decisions on the basis of high demand. However, the inevitable time lag of an office construction scheme can mean many developments come on to the market when the demand has dropped. In 1983 there were 400,000 sq. ft of available office accommodation lying vacant in Reading representing 12 per cent of the total stock of office

Block 1 continued

space. In 1984 an extra 120,000 sq. ft will come on to the market. There are few quick self-regulating mechanisms in the property market.

The politics of office development
Apart from the local property interests in the town there are few pro-office groups. The electoral and political pressure on members of Reading Borough Council has been to limit office development. Throughout the 1970s a number of attempted restrictions have been placed on developments. The Interim Strategic Office Policy, which was first produced in 1976, amended in 1977 and adopted in 1980 by both Reading and Berkshire, gave an annual guideline of 150,000 sq. ft of permitted office space. However, between 1977 and 1982 a total of 1,522,100 sq. feet of office space was given approval, an annual average of 253,683 sq. ft. The discrepancy lies in two factors:

(1) almost half of the overall 1977–1982 total (674,000 sq. ft) was approved on appeal. In other words, the local planning authority refused planning permission but the developers successfully appealed to the Department of the Environment. A case of central control over local representations. However not all appeals have been successful: out of twenty-three appeal decisions over the period 1977–1983 nine were dismissed and fourteen allowed.
(2) Berkshire County Council gave outline planning permission in June 1980 for a major office development of 344,500 sq. ft on a site it owned in King's Road, Reading. This development was in conflict with the county's own structure plan and against the wishes of the Reading Borough Council. The site was later sold to the property company MEPC for £12 million, most of which went to pay for a new county headquarters.

Note: The material presented here is taken with permission from the empirical data collected by McDonald, G. and Kidson, P. L. (1984), *An Overview of Office Development in Reading*, mimeo, Department of Geography, University of Reading.

however, new fast-growing sectors – high-tech manufacturing, producer services, public services and consumer services – have compensated for the resulting employment loss, at least in aggregate terms. Two points, however, are of note. First, while labour shedding has chiefly affected blue-collar jobs, the increase in labour demand has been relatively strongest for white-collar, predominantly female, clerical and service workers. In the high-tech industries the demand has been for highly skilled labour. Thus, even in this area of growth employment opportunities for unskilled workers are limited. Second, much of the media and political hype concerning growth along the M4 in general and the Central Berkshire section in particular has drawn attention to its private sector nature. If market forces are allowed to develop, some of the pundits argue, then the example of Central Berkshire shows what can be achieved. It is important to bear in mind, however, that most of the growth in this region has benefited from public sector spending. On the one hand there has been the major transport investments of the M4, Heathrow airport and the 125 high-speed train service. On the other hand the area has received a disproportionate share of the high level of national defence contract expenditure. The nature of this expenditure – its concentration on high-tech, the underwriting of research costs, the payment of costs plus guaranteed high profits – has been a major boost to the local economy. Almost all the high-tech firms in Central Berkshire have major defence contracts ranging from armaments manufacture to the construction of computer software for guided missile systems. The high level and type of this expenditure is one reason for the relative resilience of the local economy during the recent recession. The private sector in Central Berkshire has been underwritten by major public sector investment and public spending has been a major factor behind private growth.

Finally there are the planning applications for new private housing schemes submitted by builder-developers. In the last quarter of 1981 there were 0.26 applications for new dwellings per 100 hectares in Central Berkshire compared to 0.12 for the whole of England. The developers are seeking to meet a perceived demand for owner-occupied dwellings in Central Berkshire. This growing demand comes from a declining household size and a growing, relatively affluent local population. Central Berkshire's rate of unemployment has never been more than two-thirds of the UK average. The number of households in central Berkshire has grown from 118,210 in 1976 to 127,722 in 1980. It is not only the local economy that has provided the basis for increased housing demand. With the opening of the M4 and high-speed trains to London Central Berkshire has witnessed the development of commuting links along the growth corridor into London. According to a survey conducted by Berkshire County Council 6,900 people daily travelled to work to the Heathrow area and over 3,000 of them work at the airport alone.

The development pressure for housing takes a different form from the other two mentioned. It is more dispersed throughout the district. Moreover, while a case can be and has been made by local planning authorities for allowing employment-generating developments such as offices and factories, a case strengthened by increasing unemployment, the same cannot be said for housing. The negative impacts on the environment and local communities are more visible and the conflict between builder-developers and the local population that much more direct. Conflicts of interest between builder-developers and community groups are further heightened by the institutional purchase afforded to builder-developer participation and public participation in the planning system. In the case of housing then the conflicts are more clearly articulated and given institutional expression. It is this type of development pressure, its expression in the planning system and its effect on local community groups that is considered in this book.

The planning response

National level The planning response, especially at central government level, has been to facilitate this development. Thus, at the national and regional level Central Berkshire has been identified in a series of reports – *The South-East Study* (1964), *A Strategy for the South-East* (1967) and *The Strategic Plan for the South-East* (1970) – and endorsed by central government as an area suitable for growth. Although the exact growth figures have been revised downwards, because the birth rate has dropped and the economy has faltered, the general assumption of Central Berkshire as a designated growth area has continued to be the basis of central government thinking. Thus the review of the Strategic Plan, published in 1976, saw further growth in the area as inevitable, and in 1980 the letter from the Secretary of State to the chairman of the Standing Conference confirmed that it was the government's objective to provide facilities for further development in such growth areas.

Local level The major planning response at the local level has been the preparation of the structure plan. In the case of Central Berkshire the overall context was one of heavy development pressure and a central government committed to maintaining the growth in the area. However, there are powerful no-growth lobbies in the district which want to restrict and reduce growth. The structure plan was thus drawn up against a background of the dual competing pressures of pro- and anti-growth. In Central Berkshire the structure plan was started in 1974 and submitted in 1978. As the legislation demands there was a public participation exercise in so far as the preferred strategies were advertised, exhibited and discussed in the local media and at a public meeting held in 1979. In the structure plan three strategies were outlined:

(a) limited growth leading to a population in the mid-1980s of between 420,650 and 436,500;
(b) restricted growth (population between 399,600 and 416,200);
(c) high growth (population between 471,900 and 491,900).

TABLE 2.4: Summary of advantages and disadvantages of the alternative strategies

Key issue	*Restricted growth*	*Limited growth*	*High growth*
(1) Meeting regional needs	Not consistent with regional proposals for the area. Secretary of State unlikely to countenance revocation of residential commitments. The restrictive employment policies required unlikely to accord with the government's industrial strategy or with the regional planning strategy.	Broadly consistent with regional proposals for the *first part* of the plan period.	Not consistent with regional advice for a neutral, cautious approach to residential growth in the first part of the plan period.
(2) Central Berkshire's housing and employment needs	Inadequate scope to tackle certain housing problems or to permit a realistic response to employment pressures and opportunities. Revocation of commitments expensive and would be wasteful in areas where infrastructure already provided or programmed.	Provides a framework for meeting housing needs until the mid-1980s and possibly longer. Would permit a realistic approach to employment pressures and opportunities, provided pressures do not increase substantially.	Growth level unnecessarily high in face of anticipated housing needs. The high level of employment growth envisaged appears likely to exceed expectations unless economic prospects alter radically.
(3) The degree to which growth can be balanced with conservation and improvement of the environment	Ostensible environmental benefits would prove illusory if growth pressures returned and planning permissions were granted on appeal.	The strategy would help to contain pressures on the environment and there would be some benefits from development in Reading town centre.	Office policies could be relaxed to allow development on cleared and derelict town centre sites, but this might create more pressures and environmental problems.

TABLE 2.4: Continued

(4) The extent to which adequate transport facilities can be provided	Due to the severity of certain existing problems any benefits of a restricted growth level would be of marginal significance.	Existing transport problems unlikely to be overcome but could be contained.	Unlikely that transport problems, particularly on the highway network, could be contained.
(5) Provision of community facilities and services	No conclusive evidence that significant financial savings could be made.	Limited local authority resources unlikely to allow significant increases in standards. Precise impact of growth will depend on rate of take-up of residential commitments particularly in the first five years.	Service standards likely to decline unless more resources become available.
(6) How the plan should cope with unforeseen circumstances	Strategy not robust as difficult to implement and inflexible in the face of any increased growth pressures.	Not robust in the longer term in face of growth pressures which would develop from improved economic conditions.	An extremely high growth rate to attain over the whole of the plan period. Strategy very vulnerable if recession and economic uncertainty continue.

Source: Berkshire County Council, 1977

The consultation document of the structure plan concluded that only the limited growth strategy provided the realistic solution to the dual and contradictory demands for growth and containment. Table 2.4 summarizes the general arguments. The majority of individuals and organizations involved in the public participation scheme were against much more growth. From a sample of 1,242 electors in the area 78 per cent stated limited growth as their preferred choice while 13 per cent chose restricted growth. From the fifty-three organizations who voiced an opinion 50 per cent called for limited growth, 19 per cent for restricted growth and 31 per cent for a mixture of limited and restricted growth strategies (Berkshire County Council, 1978). Table 2.5 presents a sample of the organizations' responses.

TABLE 2.5: Local groups' responses to initial Structure Plan proposals

Organisation	*General comments*
Meadow Residents' Association (Wokingham)	Supports a national policy of conserving energy and the principle that growing communities should be socially and economically self-supporting. Ideally, therefore, there should be no growth in Central Berkshire but limited growth is supported based only on unavoidable commitments.
Hurst Village Society	Support limited growth but is concerned about the tendency for growth to lead to the coalescence of adjacent settlements. Committed to preservation and enhancement of separate communities.
Caversham Residents' Association	Prefers restricted growth but accepts that limited growth may have to be adopted.
Binfield Parish Council	Accepts limited growth but feels that restricted growth should be adopted in rural areas and villages.
Twyford Parish Council	Supports limited growth because of the need to catch up on the provision of amenities for the recent rapid growth.

Source: Berkshire County Council, 1978

The structure plan was then submitted to the DOE in June 1978, a public meeting was held in 1979 and approval was given on 14 April 1980. But a number of modifications had been made. The most important was the DOE's requirement that the county council identify land for an extra 8,000 houses on major sites over and above the initial figure of 32,000 dwellings, for development after 1982. This arose from the objection and demands raised by builders at the public meeting. An extra 8,000 dwellings meant an extra population of 20,000, the equivalent to a small-sized town. The local press quickly dubbed it 'Heseltown' after Michael Heseltine, the then Secretary of State for the Environment. In the subsequent chapters we will refer again and again to the issues thrown up by Heseltown.

The Heseltown modification was part of modifications to the structure plans of other growth areas in the South East. The Buckinghamshire County Structure Plan submitted in 1977 was amended in 1979 to make provision in South Buckinghamshire for sufficient land for housing to accommodate a further 7,500 people during the period up to 1991 and the structure plan for Surrey was amended in 1980 to make sufficient land available for a further 12,000 to 13,000 dwellings. A related modification to the Central Berkshire Structure Plan involved the DOE reducing the area to be included within the Metropolitan Green Belt. The DOE has the power to enforce central government wishes through the development control system. Most changes of land use require planning permission. This is either given or withheld by the local authority. If rejected an applicant can then appeal to the DOE who can overturn local authority decisions. Given this ultimate sanction the local authorities in Central Berkshire had little option but to accept the extra 8,000 houses. If not, they would have been faced with developers putting in planning applications which if rejected would have been repealed by the DOE. Far better, they argued, to identify sites suitable for both development and overall planning objectives than to suffer development on appeal. The questions then arose of where to put the extra houses. Another round of preferred strategies and public participation exercises was held in 1982 as the county planning department sought to find the extra land in the area.

Emerging tensions

The circumstances of agent interaction in Central Berkshire are therefore high development pressure, a land use planning system differentially accessible to public and builder participation and a government system generating political friction between its

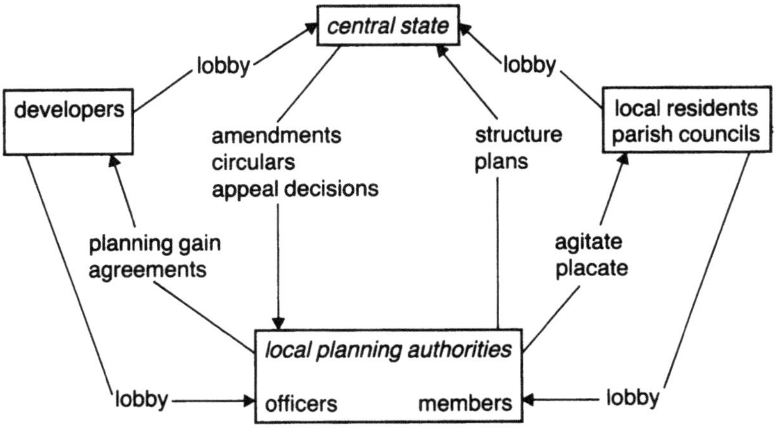

FIGURE 2.4 *Agent interactions*

different levels. Against this background a number of tensions have been emerging arising from the differing material and symbolic interests in the production of the built environment for profit, its consumption by local residents and its management by a central and local state apparatus. The state or more precisely the planning system becomes the arena for the articulation and (attempted) resolution of these conflicts. The tensions are expressed by and through the interaction of the various agents. Figure 2.4 provides a crude picture, which summarizes the major relationships. In the next four chapters we will examine the agents individually in detail looking at the nature of their links with the other agents. In the final chapter we will examine how the agents function together through a set of case studies and a more general discussion of emerging tensions.

3 The housebuilders

'Houses live and die; there is a time for building.'
(T. S. Eliot)

The production of the built environment has been a major focus of recent urban research. Harvey (1978, 1982 especially Chapters 8, 11 and 12) provides a general account while more detailed examinations are available in the proceedings of the Bartlett Summer School (BSS, 1979, 1980, 1981). In this chapter we will concentrate on the production of housing, a major element of the built environment. Housing can be produced in many ways. Cardoso and Short (1983) identify four different modes of housing production: self-produced, individual contract, institutional contract and speculative production. Here we will concentrate on the speculative production of housing for owner-occupation which since 1955 has comprised each year between 20 per cent and 30 per cent of annual output by value of the construction industry in the UK. A more general discussion of this sector is available in Ball (1983).

The production of such housing has a number of specific features which can distinguish it from other types of commodity production.

(1) Land is an essential element, which brings builder-developers into direct contact and sometimes conflict with landowners and with the whole system of land use planning both at the national and local levels.

(2) The circulation time of capital employed in the housebuilding sector is comparatively long. Money has to be spent on labour and materials and the long construction period means that capital is not realized until many months elapse and the house is sold. This means that the housebuilding sector can be very reliant on external sources of finance to bridge the gap between production and realization.

(3) The final product is expensive. The immediate purchase of

a dwelling is beyond the reach of all but the wealthiest of households. The speculative housebuilding industry can exist only through some mechanism which enables households to purchase the dwelling but phase the repayments over a period of years. Bank loans and building society mortgages fulfil this role but this means that the health of the housebuilding sector is crucially dependent on these sources of finance.

(4) The consumption of housing requires infrastructure investments. On the one hand there is the physical infrastructure of roads, pavements, sewerage and basic services of heating and lighting necessary before housing can be adequately consumed. Often there are statutory levels laid down for such provision. On the other hand there is the social infrastructure of shops, schools and cultural facilities which may or may not have such similar statutory levels but nevertheless are important elements in housing consumption and are used in households' housing choices. If this infrastructure is provided by the builder-developer the cost is offset against either lower profit levels for builders or increased house prices. Since there are limits to each of these, builder-developers often seek to pass the infrastructure course on to public authorities.

The articulation of these different features and the resultant tensions and proposed resolutions provide major points of recurring reference throughout this chapter.

Stages and agents in residential development

We can characterize the process of housebuilding as a series of stages (Table 3.1). Although the operations are presented in sequential fashion, the real process is more complex and a housebuilder is likely to be involved at various stages of the operation on different sites. The housebuilder is dependent upon a wide range of agents and the successful housebuilder is thus very concerned with coordination and negotiation.

The foundation of success in the industry is frequently seen to be knowledge and assembly of land. As a result, this is perhaps the one operation for which housebuilders have been least happy to engage the services of outside agents. Estate agents are used but often contact with landowners is more direct. Having obtained knowledge of land, evaluation of its potential is undertaken with reference, direct or indirect, to planning authorities.

Satisfied that a site has potential, the housebuilder will have to provide the capital for purchase. Outright purchase is seldom possible and many companies must resort to various financial agencies to raise the funding.

TABLE 3.1 Stages and agents in residential development

Stages	*Agents*
Land search and assembly	Landowners Estate agencies Financial agencies Planning agencies
Development design and planning permission	Architects Planning agencies Planning authorities
Housing production	Architects Subcontractors Financial agencies Public agencies
Marketing and selling	Estate agencies Building societies/banks Advertising agencies Consumers

Having purchased the land (and here a legal input is necessary), the next stage is to draw up or finalize the details of the development design for submission to the planning authority. Outside or in-house architects may be used and they frequently take the key role in organizing and negotiating this stage of planning permission.

On obtaining permission, development may begin. Labour is usually subcontracted, materials are purchased and the operation is coordinated on site normally by a site agent or the company owner. Financial agencies are often again active in providing the capital for construction and architects must negotiate with agencies for the installation of services.

Marketing of the development may begin at a very early stage. The significance of the timing of the marketing is becoming increasingly recognized (especially during times of sluggish sales) and the trend has been for the gap between building start and sale to diminish. In this way, the housebuilders' dependence upon the financial agencies is reduced. Estate agents are widely used at the sales stage but again there has been a trend amongst the larger companies to forge more direct links with the consumer. And, as an extension of marketing, some housebuilders have found it beneficial to offer an aftercare service to encourage favourable consumer opinion towards new homes. Interaction with so many other agents means that coordination and negotiation skills are crucial elements in the success of housebuilding firms.

The sequence of housebuilding takes place within a wider socio-economic framework which conditions and constrains the whole process. Before examining these stages in turn, we will consider the recent general context of housebuilding in Britain and the more local context of contemporary Central Berkshire.

The general context

The continued existence of a speculative housebuilding sector in Britain has been predicated upon rising real incomes, the nature of government involvement and the nature of mediating mechanisms. Since 1945 the steady rise in incomes has enabled more and more people to afford owner-occupation. This effective demand has been stimulated by successive governments as part of the policy to increase owner-occupation, the one constant of recent British housing policy (Short, 1982b). Governments have given a privileged position to those mediating mechanisms, the building societies in particular, which lend money for house purchase. The overall context therefore has been one attractive to the speculative housebuilding sector.

The context of Central Berkshire

As we have already noted, Central Berkshire has been the scene of heavy development pressure. Since the Second World War, the area has been a major focus of speculative housebuilding. Bracknell was early identified in the post-war New Towns scheme and successive plans and reports highlighted the suitability of the area for development.

There has been a steady demand for owner-occupied housing. Central Berkshire has a relatively affluent population well able to afford this tenure type and the buoyant economy has attracted households from the rest of the country. The population has grown from 251,000 in 1961 to 373,300 in 1980. While 18 per cent of the population of England and Wales are in the administrative and professional socio-economic groups the proportion in Central Berkshire is 22 per cent. Moreover, with the creation of good transport links to London, Reading is less than an hour from London along the M4 motorway and just under half an hour by the 125 high-speed rail link, and this has allowed substantial commuting to London. Recent years have therefore seen increasing suburbanization from the London area which has leap-frogged across the Green Belt (see Short *et al.*, 1984). In many respects, Central Berkshire is a prosperous suburban district of the London region.

To see how housebuilders operated in the contexts outlined, we conducted a survey of housebuilders in Central Berkshire. A total of forty housebuilders currently operating in Central Berkshire were interviewed (see Appendix 1). We classified the sample into four broad categories of 'very small', 'small', 'medium' and 'large' mainly on the basis of annual dwelling completions (see Table 3.2). Their responses provide the empirical material of this chapter. A much more detailed analysis of the data is available in Fleming (1984). After outlining the company organization of our sample we then go on to discuss their operation in terms of the sequence of stages outlined in Table 3.1.

TABLE 3.2: The sample of housebuilders

Number in sample	*Classification*	*Annual dwelling completions*
10	Very small	0–4
9	Small	5–20
7	Medium	21–100
14	Large	100+

The company context

The speculative housebuilding industry is financed and organized by a vast range of private investment interests. As a simplified overview of this range it is useful to identify the following ownership categories: individuals, families, consortia (i.e. owners with no or negligible family connections), and public companies (see Table 3.3). The public limited companies (PLC) category includes both firms which are themselves PLCs and those whose parent companies are PLCs.

TABLE 3.3: Ownership of housebuilding companies

Size category	*Individual*	*Family*	*Family or consortium*	*Consortium*	*Public limited companies*
Very small	8	–	–	2	–
Small	3	6	–	–	–
Medium	–	4	–	3	–
Large	–	–	2	–	11
Total	11	10	2	5	11

The individual enterpreneurs

Only the smallest of housebuilding companies were effectively owned and managed by a single person. Often developing from self-build beginnings, the numbers of one-person housebuilders are high.

However, not all register themselves as limited companies: four in the sample described themselves as being the 'proprietors' of their operation. Such arrangements obviously have their problems – loans may be very difficult to arrange and livelihoods are endangered when credit is squeezed. The unlimited operations are usually managed by relatively unsophisticated but pioneering spirits.

In all of the individual entrepreneurs surveyed, ownership and control was in the hands of one person. In many cases, spouses acted as company secretaries but in only one firm was it apparent that the spouse took an active role in decision making. Almost invariably a single person would control or coordinate all aspects of the business – but, of course, the various agents employed must wield considerable influence.

For the small housebuilding operations, then, there is sufficient room for only one executive and, indeed, that executive often took an active part in construction as foreman, craftsman, or even labourer. Expansion from such an arrangement was often resisted for fear of losing personal power over the enterprise.

Keeping it in the family

For the expanding individual entrepreneur, the most obvious way to extend the business and to retain control is to devolve ownership and/or management to other members of the family. In many cases this has been undertaken on a father to son(s) basis. Entire control over very small to medium firms can be maintained by families but they seldom extend to large companies where much external finance and personnel is necessary. Even in the medium-sized companies which were entirely owned by the extended family, it was usually necessary to draw upon non-family personnel for many management functions.

In the small family firms, administrative duties were spread amongst the various directors in an *ad hoc* and flexible fashion. Although the term managing director was seldom openly used it did appear that one director (typically the father or eldest brother) was often dominant and took the leading role. The executive staff would become involved in most aspects of the housebuilding process and sometimes there would be a weak demarcation of duties according to site.

In the medium-sized family firms, however, management tended to operate on a much more structural basis. Since these firms often had other non-housebuilding interests, certain personnel had little direct involvement with housing. The most common demarcation for those directly involved in housebuilding was between the land search and the site work.

The adventurous entrepreneurs – the consortia

Consortia of non-family based private interests were surprisingly few. Between the family-based firms and the public limited companies, there appears to be a gap which is not easily filled. There are a number of reasons for this.

First, there is a reluctance amongst individuals to cooperate in a housebuilding venture (in a later section on land search and aquisition (p. 52), the virtual absence of communication amongst the very small and small firms will become apparent). The pride and independence of the entrepreneur and the high-risk nature of the business combine to encourage isolation. Even partnerships are rare.

Second, individuals or groups who do cooperate will tend to be ambitious for growth. As noted in the section on the dynamics of housebuilding companies (p. 281), inertia is a powerful force in housebuilding. However, should expansion be successful, the company may be encouraged to go the full distance to become public.

In sophistication the consortia stand apart from the individual and family-run firms. Consortia personnel tend to be drawn from within and without the housebuilding industry, but commonly the executives are experienced, specialized and ambitious. On occasions, however, the ambition is tempered by experience with other (often ailing) companies from which the personnel has been drawn.

Organizational structures in consortia tend to be much more clearly defined. In the two very small consortia firms all day-to-day management was in the hands of a managing director whilst the other director(s) merely had a financial (and in one case a legal) input to the company. In the medium-sized consortia companies, and particularly where there are other non-housebuilding interests, demarcation of duties is often very rigid.

Gone public

Of the thirteen large companies surveyed, eleven (85 per cent) were either public limited companies in their own right or subsidiaries of such. This proportion is high and may result from the peculiar nature of the study area as a growth zone – an area of growth is bound to attract the major housebuilders.

The public companies may manage their housebuilding operations on a centralized or decentralized basis. Often a company will establish subsidiary limited companies to operate in particular regions. In such instances, very considerable autonomy is granted to the subsidiary companies. In other circumstances, all decisions will emanate from one source. And their are yet other cases where a loose form of decentralization is encountered: for one major company, three regional groups were established with considerable devolved power to take decisions about particular sites; however, all the planning decisions were made from a central headquarters. Table 3.4 gives some impression of the allocation of administrative responsibilities within housebuilding divisions.

TABLE 3.4: Administrative staffing of a major housebuilding subsidiary company

4	land negotiators
2	surveyors
3	production managers
1	subcontracts manager
1	engineer
2	maintenance managers
1	landscape architect
1	showhouse architect

Land assembly

Speculative housebuilding is a high-risk enterprise. Land, a relatively scarce and high-priced commodity, must be acquired and serviced over a period of time before the dwellings can be constructed and sold in fluctuating market conditions. The assembly of sufficient quantities of land at the right price is therefore a major concern of housebuilding companies. In this section we will examine the land assembly procedures of our sample of housebuilders.

Block 2 The very small housebuilder

'I'm a disillusioned perfectionist.'

Mr X is the classic solid, deferential, small housebuilder of limited ambition. Living in his own unflamboyant, spacious, self-built detached house in a bland suburb of Reading, he is every inch a self-made man, very conscious of his own upward social mobility.

As a 26-year-old jobbing contractor, he built his own house (not his current abode) thirteen years ago. That was to be the beginnings of his housebuilding activity. His contracting work was quickly reduced but persists to this day and is now supplemented by some small commercial development. Housebuilding, however, is regarded as his main earning element even though the early 1980s recession has caused him to redirect much of his personal capital to support his limited company status. Output varies between one and three dwellings annually (he likes to complete one job before embarking upon another), but his attention to detail is constant. Much of his working practices are geared to ensure quality – he personally designs his traditionally built dwellings and he works alongside his subcontracted workforce which consists of older workmen paid on a daily basis. Young workmen are avoided because he considers they are inadequately trained and piece-work payment is avoided because he thinks this results in rushed and shoddy work. But ultimately he is frustrated by the consumer's lack of appreciation of his product. His eventual product is bland – almost exclusively £50–60,000 four-bedroomed detached houses.

Despite claims that he would like to expand, he is doomed to be a very small housebuilder because of his practices. The land search is for land with planning permission obtained (at relatively high prices) solely from estate agents and in response to advertisements. Attempts to expand by the purchase of a builders' yard were frustrated by the predictable higher bid of a bigger local builder. In any case, his ambitions for expansion are tempered by his satisfaction in being his 'own governor'.

Land search

Financial strength is a useful but not overriding factor in the acquisition of prime land – much more important is early knowledge and early securing of the resource by outright purchase, option agreement or conditional contract. Previous work such as that by Drewett (1973), Craven (1970) and Bather (1973) has highlighted some of the main methods of acquisition of knowledge of land but they have tended to ignore the different strategies and tactics operated by different developers.

Faced with an enormous range of options, one of the earliest decisions which must be taken is the regional limitation of the land search. Although a housebuilder may attempt to define his area of operation quite tightly, it is inevitable that he will obtain information about sites which are located outside his traditional area. If the potential seems great, he may well be tempted to follow up such sites. At first these follow-ups may prove abortive but, invariably, they will lead to a more detailed knowledge of a wider area and may ultimately lead to expansion of operational areas. To what extent these distant temptations are encountered purely by chance and to what extent they are deliberately forwarded by land agents to promising developers or to a range of developers during hard times is an interesting if complex question. Any house buyer will have recognized the attempts of estate agents to interest potential buyers in areas about which information was not initially requested. This is merely a way of pushing one's own product and encouraging favoured developers.

Very similar points may be made about the size of site sought by a developer. Site size desired is fairly closely related to a developer's annual number of completions (and more fundamentally with the smaller-sized developers to loan facilities and capital available). Generally, the most favoured sites of the very small firms are the single-dwelling luxury plots; small firms prefer sites designed for between five and ten dwellings, the medium-sized firms prefer sites for thirty to fifty dwellings; while the large operators frequently favour sites of around 100 dwellings.

Preferred sites and acceptable sites are of course quite different concepts and in difficult times developers are often forced or encouraged to build on sites which are very much smaller than preferred. It is much more difficult to consider sites which are larger than normal limits because financial restrictions impose very definite prohibitions.

Strategies of the land search

Two main strategies can be identified:

(1) the saturation or systematic approach;
(2) the opportunist or selective pickings approach.

There is some overlap between these approaches and both can be operated simultaneously but it is none the less very useful to distinguish between them.

The *saturation strategy* identified by Drewett (1973) involves a comprehensive survey of an entire area. One such developer in this sample began by systematically studying the latest 1:10,000 Ordnance Survey maps for an area to identify possible sites. Consultation of the local development plans was supplemented by reference to the plotting sheets which are often constructed to identify sites with planning permission or for which planning permission had been sought. Particularly good relations with certain authorities enabled access to the individual planning application files from which information could be gleaned about the names of the owners of the land and reactions to past proposals for development. Although this may seem very systematic, the wealth of experience of the land searcher plays an enormous role in the following up of certain sites. The approach is thorough but very time-consuming.

Clearly this procedure is most suited to large companies with sufficient output and staffing to justify an extended land search. The processes of land search and subsequent negotiation are of course closely allied, and to precisely identify numbers involved in land search activities of a company is difficult. None the less, of the fourteen large companies no less than ten (71 per cent) employed at least one person full-time in land search and negotiation. Such full-time commitment was evident in only one medium-sized company and in none of the smaller firms. This is indicative of the extent to which information about land is almost subconsciously acquired by the local and smaller companies. To the large firm with substantial areas to cover and new areas to explore, the specialization and the commitment of at least one employee is vital.

The saturation strategy was thought by Craven (1970) to be on the decline. Its initial popularity stems from the early 1960s when planning policies were so loosely defined that developers considered it almost to be part of their job to bring land to the attention of planning authorities. With the passage of years, however, and the advent of structure planning, the planning authorities began to indicate more precisely where development might be appropriate. In effect, the land search could be confined to a much smaller area and the saturation approach became less appropriate – instead a 'head-hunt' (the identification and acquisition of the pre-defined prime sites) become more relevant.

It would appear that in the 1970s, the saturation strategy was only adopted by those developers who wished to identify land ahead of the publication of local development plans. One of the main purposes of the saturation strategy is to acquire land for which no great demand is apparent. The landowner of a promising site is approached directly by the developer so that the land may be secured at a relatively low cost. Identification of development sites prior to official designation should therefore provide a wide choice and enable cheaper acquisition. The use of option agreements or conditional contracts rather than outright purchase will help defray the economic cost of a misjudged acquisition.

From this survey of developers in Central Berkshire, it was

apparent that the saturation strategy had undergone something of a revival. Five of the large companies had adopted an approximation of this strategy in an attempt to identify the land for 'Heseltown'. Although Berkshire County Council produced a document in 1982 which shortlisted (though not irrevocably) potential sites, many developers adopted a saturation policy in an attempt to acquire the land prior to the publication of the results of the Berkshire County Council study.

The structure plan modification, then, suddenly provided a much more fluid planning environment and marked, from the land searcher's point-of-view, a movement (though short-lived) away from the restrictive planning policies of the last decade.

One interesting aspect of the revival of the saturation strategy was the use by developers of outside architectural/planning consultants. They could provide a detailed local knowledge backed by some of the national developers but, more interestingly, they could be used by consortia of developers thus encouraging a degree of cooperation between the major developers.

The saturation strategy is used only by some of the large companies. For the medium and smaller-sized developers, the approach is hard to justify for a number of reasons: they are generally not seeking the large sites which are best identified by this approach; they lack the resources to undertake or commission such surveys; and as local developers, their knowledge will already be very extensive though not published or verbalized in any systematic form.

The *opportunist strategy* For all developers, and as the exclusive approach of the smaller operators, the opportunist strategy is vital. The approach is opportunist in the sense that the developer reacts to land opportunities which come to his attention. But the role of the developer should not be interpreted as being entirely passive: considerable thought and effort may be invested in attracting and creating these opportunities. Superficially, it may appear possible to identify distinct passive and active, receptive and aggressive styles but often this distinction cannot easily be made. Like the opportunist goal scorer in a football team, the astute developer will await the good pass and dispatch it as necessary. Receiving the pass is seldom luck: the developer has established which players provide the best opportunities and carefully positions himself so that he appears to be the obvious choice to receive the opportunity and, indeed, difficult to outflank. Positioning is the result of establishing a wide range of business and perhaps more importantly para-business or social contacts.

Sources of information

For the purpose of the builder-developer interview a number of sources from which information could be obtained about land and acquisition sources were identified and respondents asked to select from a predetermined scale the frequency with which they used these sources (see Tables 3.5 and 3.6). Much, of course, depended upon the respondent's own definitions of 'always', 'frequently', 'occasionally', 'rarely' and 'never', but in practice definitions appeared to be broadly standard. A ranking procedure was considered but would have been less easy to administer and in some ways more difficult to interpret. Although some developers could have given more detailed and quantitative information about their information sources, such information could not have been collected comprehensively and so was not specifically requested. One final point worthy of mention is that for the large companies especially, interests in sites of very different sizes often means differential use of information sources and sources of supply – to rate these on one scale can be slightly misleading. Attention will be drawn to such instances in the course of the discussion.

The interview identified seven major sources of information about land: estate agents, media advertisements, direct contacts with private landowners, other builder-developers, non-developer companies, solicitors and other professional contacts, and the study of planning documents. This list is not comprehensive but it does encompass the major sources.

Cycles of searching

The function of the land search is to build and to maintain a bank of developable land so that the company can operate consistently at a desired pace. Since land opportunities are always coming on to the market, it is preferable to operate a steady policy of land search. Slack and busy periods may occur as sequences of opportunities are realized or lost but the aim is to maintain a steady flow.

One landfinder and negotiator of a major developer considered that between 5 and 10 per cent of his time was taken up by the actual land search and a further 90 per cent in investigating and negotiating the sites which he had found. For various reasons the uptake of opportunities is low: the same land negotiator was currently looking at ten sites, expected to try to obtain four but considered he would be lucky to obtain one. To pick up five sites in a year would be considered an exceptional achievement,

TABLE 3.5: Sources of land information: number of firms claiming to receive information from various sources

	F	O	R	N	F	O	R	N	F	O	R	N
	Source of information											
Size of Company	Estate agents				Own advertisement				Responding to advertisement			
Very small (10)	7	1	1	1	–	–	–	10	2	3	1	4
Small (9)	7	1	–	1	–	–	2	7	–	3	3	3
Medium (7)	5	1	1	–	–	1	–	6	–	3	3	1
Large (14)	10	1	2	1	3	3	3	5	9	1	2	2
Total (40)	29	4	4	3	3	4	5	28	11	10	9	10
	Approach by landowners				Approach to landowners				Other housebuilders			
Very small	1	1	7	1	4	3	2	1	–	1	1	8
Small	1	3	3	1	1	3	4	–	–	1	2	6
Medium	1	3	3	–	2	4	1	–	–	5	1	1
Large	7	2	3	2	4	4	3	3	4	8	1	1
Total	10	9	16	4	11	14	10	4	4	15	5	16
	Other companies				The professions				Plan Yes		Study No	
Very small	–	–	–	10	–	3	–	7	1		9	
Small	–	–	3	6	–	2	2	5	5		4	
Medium	–	1	2	4	–	2	2	3	6		1	
Large	2	7	2	3	1	3	4	6	12		2	
Total	2	8	7	23	1	10	8	20	24		16	

(F = frequently, O = occasionally, R = rarely, N = never)

although in the course of the year he might consider 100 sites or more. This is not considered unusual for a large, competitive and ambitious company with a policy of growth. For the smaller developer the ratio of uptake to investigation might not be smaller but the amount of detailed work going into the search and investigation would be considerably less.

Ease in obtaining land is generally related to its price. Land bought direct from a private landowner may require considerable search and negotiation but its price should be reasonably low. In contrast, land can be easily obtained at auction but at a price. It is therefore possible for developers with a sizeable land bank to spend more time in obtaining the cheaper land while the developer anxiously trying to acquire his next site is often forced to do so

TABLE 3.6: Sources of land purchase: number of firms claiming to use different sources

Size of company	F	O	R	N	F	O	R	N	F	O	R	N
	Source of information											
	Estate agents				Private landowners				Other housebuilders			
Very small (10)	4	3	2	1	5	5	–	–	–	–	2	8
Small (9)	6	2	–	1	3	3	1	2	–	1	1	7
Medium (7)	3	1	3	–	2	4	1	–	–	3	–	4
Large (14)	5	4	4	1	5	4	4	1	4	6	3	1
Total (40)	18	10	9	3	15	16	6	3	4	10	6	20
	Other companies				Receivers				By auction			
Very small	–	–	1	9	–	–	–	10	2	2	2	4
Small	–	–	2	5	–	2	–	5	–	3	2	4
Medium	–	1	2	4	–	2	2	3	–	1	2	4
Large	–	6	6	2	–	4	3	7	1	2	7	4
Total	–	7	11	20	–	8	5	25	3	8	13	16

	Local authorities				
	F	O	R	N	N/A
Very small	–	–	5	1	4
Small	–	1	1	2	4
Medium	–	–	3	3	1
Large	6	1	3	1	3
Total	6	2	12	7	12

(F = frequently, O = occasionally, R = rarely, N = never, N/A = not available)

by the quick but more expensive methods. A spiral may thereby develop which can retard or spur a company's growth.

One might expect companies which are seeking to expand to adopt identifiably different land search strategies in order to obtain sizeable and cheap sites necessary for growth. Since so few of the very small to medium firms and so many of the large firms are expanding, it is difficult to determine how the methods of the expansionist firms compare to their stable or stagnant counterparts. It is reasonable to assume that growth necessitates a wider land search procedure but the critical differences relate to financial backing and, in certain cases, to a willingness to consider, and astuteness in assessment of, land without planning permission.

The emphasis placed upon different sources of land varies with the size of housebuilder. These are most effectively differentiated by three of the most important sources: the estate agent, the private landlord and other builder-developers.

Typically, the very small builders rely almost exclusively upon purchases direct from the private landowner, via the estate agent and (increasingly) from the local authority in that order of importance. They are more likely to approach than to be approached by the private landowner, they will respond to adverts in the local media but they will hardly ever receive or act upon information from another developer.

The small builders, in contrast, rely very heavily upon the estate agent, less on the private landowner and even less on the local authority. Their land search will not cover a great range of sources.

The medium-sized builders rely about equally upon estate agents and the private landowner for actual purchases but they do have a reasonably diverse source of supply. If they operate on the large sites they become increasingly interested in inter-developer dealings.

Standing apart from the rest are the large builders, often with their specialist land search department. A very wide range of information avenues are explored and this is reflected in their very diverse range of sources of purchase. Other builder-developers, local authorities, estate agents, and private landowners are of about equal importance in terms of numbers of purchases but it tends to be other builder developers and the private landowner who provide the largest sites and it is the last – the private landowner – who provide the choice pickings.

Of all sources, it is the use of the private landowner which best helps to distinguish the style and approach of the housebuilder. Generally, dealings with a private landowner imply that the land may not necessarily obtain planning permission and that the relevant developer has made the first approach. All this suggests that a particular developer is prepared to take risks and to invest substantial resources in obtaining access to such land: these are the land-finding housebuilders.

In contrast there are other housebuilders who rely almost entirely upon second-hand or third-hand information and buy their land through agencies or others with development interests. This land is more expensive and generally it is found that companies following this type of policy invest greater resources in marketing their ultimate product. These are the marketing housebuilders and they appear to be increasing in importance as the housebuilding industry slowly transforms from an undermanaged enterprise to a more sophisticated operation.

Block 3 The small housebuilder

'I've done it all before'

From being a rising bright young thing, this small-time housebuilder settled for the domesticity of safe development. Established in the 1920s, his old family firm was completing around 150 homes per year in the early 1970s. But in the slump of 1973/4, the firm 'caught a cold' and it was thought best to lay it to rest two years later.

After a lapse of a few years, two brothers of the family formed a partnership to undertake a modest housebuilding enterprise. With a turnover of less than £1 million, completion rates varied from ten to 20 even though forty could have been produced if desired. Despite all the trappings of a small stagnant enterprise, this was in fact a firm of sophistication.

The land search covered a very wide area – a useful relic of the old family firm. Unlike most small housebuilders, there was little reliance upon the estate agent for land: private initiatives to landowners were deemed much more effective and profitable. Capital and sales came through traditional sources – banks and estate agents – but the outstanding feature of the firm was its approach to the planning system.

Planning was highly respected and the enterprise was geared very much around manipulating the system (albeit with style and taste). Based on a philosophy which interpreted the committee's role as representatives of existing residents, and of the planning officers' role as representatives of the new purchasers, the *basis* of all negotiations was with the latter. Often, but only with the planning officers' consent, approaches would be made directly to parish and district councillors to persuade them of the quality of the proposed development. Of course, modifications and concessions would be considered but always the bottom line was to extract a substantial profit from a well thought-out development. Professionalism, if necessary through the contracting of professionals, was the mode of operation; a comfortable and secure living, the outcome.

Land banking

Land is a basic requirement for builders. To ensure continued production builders often hold supplies of land greater than that currently used. Smyth (1982) describes a 'land bank cycle' whereby land may go through phases from acquisition through planning permission to being built upon. Land may be 'held' by a developer in a number of ways. Most simply, the land may be owned outright. For a simple developable site this may be the most economic and convenient mode of acquisition. However, should the site not obtain planning permission, the capital returns will be low. If the site takes some time to develop and capital has been borrowed, interest rates will have to be paid. Housebuilders buying land outright are therefore under pressure to process the site and sell the dwellings rapidly.

Ultimately, housebuilders must buy the land upon which they build. None the less, payment may be delayed (to avoid excess payment of interest) or conditional (to avoid the purchase of an undesirable or undevelopment site). To hold land without the necessity of outright purchase, two arrangements are encountered.

First, there is the *option agreement*. An option agreement involves a once-for-all payment to obtain the sole purchasing rights for a site for a specified period of time. Agreements vary in detail but in nearly all cases the vendor is committed to sell should the developer wish to purchase – the developer is not committed to buy. The eventual sale price is usually fixed at the existing market price. Private landowners are often tempted by the initial payments (of a few thousand pounds for a substantial site). The developer can then direct resources to a site with a certainty that should the site prove suitable, it can be bought.

Second, there is the *conditional contract*. It is a stricter arrangement whereby both vendor and purchaser are committed to a transaction if certain conditions (usually the obtaining of planning permission) are satisfied. Because of the great commitment required by conditional contracts, developers favour option agreements.

Option agreements and conditional contracts are utilized most by the large housebuilders. In Central Berkshire, these arrangements have long been in use. It would appear that of the potential 'Heseltown' sites all but two had been secured by developers before the final announcement of the allocation. The owners of one of the unsecured sites had long resisted attempts to secure their land, confident in the knowledge that their land would eventually obtain residential zoning. They felt they should hold out until the land was eventually earmarked so that they could obtain the best possible price. Indeed, it is thought that most of the 'Heseltown' sites had been secured prior to the 8,000-houses modification – developers, at least, were in no doubt that Central Berkshire was to be an area of growth. For those developers with option agreements on sites which were eventually rejected as 'Heseltown' sites, losses were minimal. Indeed, they may well maintain an interest in these sites since, if Central Berkshire continues to be a designated growth area, it is likely that they will eventually be allocated. Trading of options is often undertaken – it is known that an option on one of the large Central Berkshire sites was bought and sold by a small builder on a purely speculative basis.

The concept of the land bank is therefore complex. Though not owned, land held through options or contracts forms a special type of land bank. Yet many large developers have little idea of the amount of land they hold through such arrangements. Large sites are obtained most easily 'en bloc' by the option agreement. Few but the largest developers therefore compete. However, the sites which are acquired by this method are often very much larger than those which they might wish to develop. None the less, they are cheap to acquire and do give the purchaser a degree of monopoly control over the site – that company can select the best plots and exercise some control in selecting neighbouring developers. The nature of the option agreement, therefore, goes some way to explain the high degree of inter-builder land sales amongst the large developers and the need for their cooperation. It is in their interests to communicate as much as to compete.

Whereas option agreements and conditional contracts stake a developer's interest in a site, *outright purchase* requires the ultimate capital commitment. This may be a liability if the site does not receive planning permission (hence land is usually only bought at developable market value if it already has a planning consent), but in times of rapidly rising land prices it is of very considerable benefit.

If the inflation of land prices between land purchases and house sale exceeds the general inflation rate, the housebuilder will profit. The profit may of course be severely reduced by the need to pay off interest rate charges if the land was purchased with borrowed money. The general inflation rate, housing and land prices and interest rates have varied considerably over the last decade. Housebuilders' attitudes to land banks have fluctuated according to the state of the market; with recent high interest rates and the general economic recession and high land prices, the trend has been to label banks as a thing of the past (Joint Land Requirements Committee, 1983). Recently, land assembly has been most effectively undertaken via the more flexible and less risky option arrangements.

Given that the land bank is a complex concept, it is very difficult to obtain an overall picture of a company's interests in land. To gain an impression of each company's stock of owned land, the following question was asked: 'At your future expected rate of building, how long will your currently owned land meet your requirements?' The responses are displayed in Table 3.7.

Generally, the larger the company the longer the land bank was expected to last. This reflects the desire of the larger companies to keep conservatively large land banks to ensure a steady rate of construction, and their ability to secure and service large land banks. It also reflects the fact that in general large companies

TABLE 3.7: Housebuilders' perception of the adequacy of the size of their land bank

	N/A	*Adequate*	*Too much*	*Too little*
Very small	2	2	1	5
Small	1	2	1	5
Medium	–	4	–	3
Large	–	10	–	4
Total	3	18	2	17

build on relatively large sites and that these take longer to progress through the land bank cycle. In contrast, the smaller firms can better cope with shorter duration land banks.

Different sites move through the land bank cycle at different rates. For any housebuilding firm, certain of its sites will be expected to take longer than others so there is no regular or ordered progression through the cycle. Three examples should suffice to demonstrate this. A medium-sized builder owned and farmed twenty acres of agricultural land. This land was included in the land bank even though the firm realized that the land might not receive planning permission for two decades. The land would make a very slow progression through the cycle. A small builder had his land bank plans upset because, unexpectedly, he had found it necessary to take a site to appeal to obtain higher densities. The appeal process effectively put a temporary stop (of probably six months) on the progress of this site through the cycle. Finally, there is a trend amongst the larger companies to seek out small quickly developable 'top-up' sites. The recession and perhaps planning delays have encouraged some large companies to try to maintain a steady and sustained output through the use of small, if expensive sites, ripe for development.

Perceived adequacy of the Land Bank

The land availability debate has attracted increasing attention throughout the 1970s and 1980s. Housebuilders' Federation (HBF) pressure at the national and local level has urged more land release, and in Central Berkshire such pressure resulted in the 'Heseltown' allocation. But what do the housebuilders of a growth area think of the size of their land banks?

Using the simple categories of 'adequate', 'too much' and 'too little', it is apparent from Table 3.7 that very few of those interviewed considered that they possessed too large a land bank – selling is a simple solution to such problems. Of the two who considered that they owned too much, one was the aforementioned small builder whose land bank cycle had been upset by his decision to take a site to appeal: temporarily, he had too much land. The other was a very small firm which dabbled in housebuilding and, at the admission of its director, 'could not get it right' – buying land at the wrong time and failing to sell dwellings promptly.

The claim of most developers to hold an 'adequate' land bank needs clarification. The holding of an adequate supply involves a compromise between capital outlay and future building rate security. While land may be scarce it is generally conceded that it is company finances which preclude the possession of a larger land bank. (The problem becomes even more complex when it is realized that the scarcity of land is a contributory factor to its high price thus preventing developers from enlarging their land banks.) It is notable therefore that while 70 per cent of the large firms claimed to be content with their land bank, only 25 per cent of the small and very small firms professed similar satisfaction.

Of the small and very small firms who claimed to possess too small a land bank, only one attributed this solely to the lack of sites with suitable planning permission. Generally, it was admitted that suitable sites could be purchased – but at a price. And the price for many suitable sites was frequently thought to be excessive.

Measured in terms of area or duration, it is obvious that different sizes of developers consider different sizes of land bank most appropriate to their requirements. For the small and very small developers, two or perhaps three years is generally seen as a comfortable size of land bank. However, uncertain economic conditions and the necessary capital often prohibit the luxury of such land banks. The larger developers tend to want to hold land stocks of between three and five years. In fact, their stocks tend to be even larger as a result of their background option agreements.

Distinction is often made between profits made by housebuilders on construction turnover and upon land. Smyth (1982) has indicated that to profit substantially from the land element, it is necessary for a company to have a land bank of approaching ten years. He also noted, however, that throughout the 1970s there was a shift from profit making on the land element towards profit making on increased turnover. The results of the Central Berkshire survey suggest that in the early 1980s, in an area of growth, there is even greater emphasis upon turnover profits. Companies with ambitions of increased turnover are bound to be attracted to an area of relative growth where demand is high.

Planning status of land when purchased

To examine further the distinction between housebuilders who derive substantial profits from increases in land prices and those who rely almost entirely upon construction profits, it is illuminating to analyse the planning status of the land which developers purchase.

Land may be bought with 'detailed', 'outline', or without planning permission. The last category, however, is ambiguous. Developers and landowners are not naive of the workings of the planning system. Experience or direct questioning of planning officers may give an impression of site potential. For small sites, the 'infilling' or 'rounding-off' policies are often explicit enough to judge potential. Borderline cases may require further investigation and/or discussions with planning officers. In other instances permission to develop might seem a remote possibility, yet purchase is undertaken. Purchase price generally reflects the planning status of the land: the 'probable' development site is likely to obtain full development land value whereas the truly speculative site might be obtained at little above its agricultural land value. Hence, substantial profits may be made by obtaining planning permission for a site.

Most developers seek to buy land with outline planning permission. Of those interviewed, 88 per cent claimed to buy land with outline consent 'frequently' and only two (5 per cent) claimed never to have purchased land with outline permission. However, this is a little misleading because it implies that landowners secure planning permission for their land before sale. In practice, either formally through options and contracts or more informally, there are often strong commitments between landowners seeking to obtain permission for their land and interested developers.

The 'planning profit' potential from sites with outline planning permission is small. Generally, the vendor will sell at the going market rate for designated land and only if the purchaser can modify the existing planning permission will (s)he be able to extract further planning profit. (To some developers this is known as 'planning gain' – contrasting with the more commonly understood planning gain whereby the planning authority manages to extract some contribution from the developer in return for permission to develop a site.) The most common method of achieving this first planning gain is to obtain permission for a greater density of dwellings on a site. By careful design and layout, one small housebuilder recently extended outline permission from five dwellings to detailed permission for eight, thus greatly enhancing his profits on the site.

Very few developers wish to buy land with existing detailed permission, although few manage to avoid this circumstance altogether. Detailed permission, usually obtained by another developer rather than the original landowner, is generally too restrictive and nearly all will reapply to obtain a detailed permission which is more suited to their style of construction. In fact, many developers look upon land with detailed consent with suspicion – they will want to know why an applicant (usually another developer) will have gone to the trouble of obtaining full permission and then failed to capitalize upon this effort. It also tends to be more difficult to extract further planning profit from a site with detailed consent.

Just under half of the companies interviewed claimed to have completed a purchase of land which had no existing planning consent. Naturally, the purchasers considered that the odds of obtaining permission were acceptably low. In compensation for the risk, the planning profit, if obtained, would be expected to be high. No interviewee claimed this as a 'frequent' mode of purchase and it did tend to be most common amongst the smaller housebuilders. The sites bought will therefore have tended to have been small – and often subject to the structure plan rounding-off or infilling policies. Large sites are deemed too risky to purchase outright without a more tangible planning commitment.

Of those who purchased land without official consent, just over half were also willing to buy land on a highly speculative basis (i.e. without even informal planning commitments). Such land is usually obtained at very low cost and would therefore have extremely high planning profit potential. But the risks are correspondingly higher. Speculative buying of this nature was rare and undertaken only by the daring or the desperate.

The purchase of land with or without planning permission transcends size groupings. It does, however, provide a revealing distinction between the housebuilders who attempt to receive substantial profits from the land (the 'landfinders') and those who rely upon turnover (the 'constructors'). Further subdivisions of the constructors and the landfinders will be attempted in later sections: different types of constructor are best discussed in the section concerning marketing, and of landfinders in the planning section.

The philosophy of the constructors can be typified as: 'dear land is cheap land and cheap land is dear land'. Land with planning permission fetches a high price but is prime development land and should involve no marketing problems; cheap land, in contrast, will probably have no planning permission, may have no prospect of receiving any and, even if it does, may in the long-term be laden with infrastructure costs and marketing difficulties. This is a conservative low-risk outlook and forces a company to concentrate on construction and rapid marketing. The accompanying land search will be of a relatively passive nature and emphasis will be placed upon estimates of costings of sites. Using this approach and with no major external finances, small firms are effectively 'locked into' their size category. Profits will be modest but consistent and the future foreseeable. But for the major housebuilders who can buy segments of the large sites, emphasis upon construction and marketing may enable continuing growth and expansion and need not lead to stagnancy.

The style of the landfinders is much more aggressive and may be supported by varying degrees of planning awareness and may arise from confidence and strength or flamboyance and ambition. There is even some circumstantial evidence to suggest that companies threatened with liquidation sometimes attempt to find a rescue route through the purchase of hopeful sites; success may mean survival, failure often entails a rapid demise.

The production of housing

Operational areas

Contrary to most industrial activity, the construction industry in general and the housebuilding industry in particular is a nomadic enterprise. Movement is necessitated because once the land resource has been used, it is not economically viable to recycle it for a considerable length of time.

The conflicting pressure to confine activity to a relatively small area stems from several influences, the most important being the need to minimize travel costs. It was intriguing to note, however, how one builder used this travel time positively: by taking as many different routes to the operational sites as possible he used these journeys to identify potential future sites. Daily movement from headquarters to site of men, materials and plant can be expensive. Even using subcontracted labour, restrictions can be imposed because to retain a particular gang it may be necessary to offer short travel distances: one very small developer was restrained in this way because he wished to retain a group of bricklayers who disliked travelling. Moreover, as the discussion of land search and acquisition showed, the detailed knowledge of an area, including contacts with local planners and agents of the development industry, encourages a relatively small area of operation.

The size of operational area is closely related to size of housebuilder. When asked to define their areas of operation, most housebuilders spontaneously replied in terms of a mileage radius: these are presented in Table 3.8. This table indicates clearly that the larger the firm the greater its operational area will tend to be. Different sizes of housebuilders look for different sizes of site and larger areas afford greater potential and choice. The large companies could not define their operational areas in such terms and will be considered separately.

TABLE 3.8: Operational radii of different sizes of housebuilder

	Operational radius (miles)				
Size of Housebuilder	10	11–20	21–30	31–50	N/A
Very small	5	2	–	–	3
Small	1	4	3	–	1
Medium	1	–	2	1	3
Large	–	–	–	–	14
Total	7	6	5	1	21

If a graph is plotted relating a housebuilder's annual number of completions to operational radius, it is possible to identify two extremes of housebuilders (Figure 3.1). First, there is a group (A) with a relatively high operational radius in respect of number of annual completions. These firms can be characterized as of two distinct types: those whose housebuilding activities are declining and whose operational area is anachronistic and larger than might be currently necessary; and those whose housebuilding activities are increasing and realize that to achieve increased production a necessary prequisite is to expand the area of operation to increase opportunities.

Second, there is a group (B) with a relatively small operational area for the number of houses built annually. These tend to be well-established firms with a consistent output but who have a sufficiently high level of knowledge and quality of contact to enable operation within a fairly small area.

Extremely few housebuilders of medium size or less were conscious of expanding or contracting operational areas, thus reinforcing the notion that the concept of a growth cycle of a housebuilding firm is inappropriate. Territories are quickly established and adhered to, and, to a certain extent, reflect company building programmes.

None of the large companies considered operational areas in

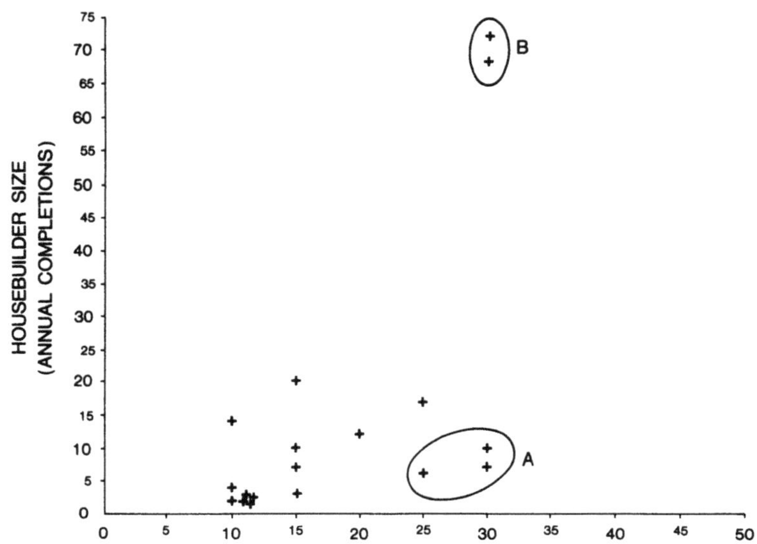

FIGURE 3.1 *Operational radii of housebuilders in Central Berkshire*

terms of radii from headquarters. Instead they operated in regions of interest, but to accomplish this it was often necessary to establish separate divisions or even limited companies within their own right.

Those companies which operate over a very wide area tend to have a greater number of subsidiary companies or division. These subsidiaries are usually established when geographical expansion is considered or confirmed by success in an area. Such expansion is still proceeding but is not always successful – one Eastern England-based firm established a Central Berkshire-based division in the hope of expansion but the move was soon abandoned because of problems of communication and poor site selection.

The autonomy given to separate divisions or companies varies enormously. Some are highly autonomous with their own budgets and land search and planning departments, while others are merely site monitors whose existence is warranted by their facilitating but not initiating actions.

Central Berkshire as an operating area

Throughout the 1970s and the 1980s, Central Berkshire has been a very attractive area for housebuilders. Those developers of medium size confine most of their activities to this administrative region. This is convenient in obtaining a thorough knowledge of an area and in establishing links with a few local planning authorities – an undoubted key to success. The existence of the growth areas has not attracted many developers of this size to expand into the area at the expense of their own local spheres of interest. The only firms of this size which have really seized upon the potential of Central Berkshire as their principal operating area are a very few housebuilders based on the London side of the Green Belt. Such a location of headquarters is anomalous because very little development is now undertaken in these areas. Inertia has prevented them from moving, but to continue, their sites must be located at some distance from their headquarters.

For the medium and smaller-sized firms, then, the concept of Central Berkshire as a growth area is not of vital significance. Those developers based at some distance from Central Berkshire have made little attempt to extend into the area; nor have developers of this size relocated to the area to take advantage of its opportunities.

For the large housebuilders, a very different pattern emerges. Only one firm operated solely within Central Berkshire, whilst only another three had regional offices with a degree of policy power (as opposed to site or sales offices) in the structure plan area. However, this lack of devolution to Central Berkshire is not altogether surprising: of the remaining ten companies, five were highly centralized operations, and of the remainder three had divisional offices within fifty miles of Reading and two had only Central Berkshire interests.

For the thirteen large companies, the first levels of control of Central Berkshire operations were located as follows:

Within Central Berkshire	4
London	6
Within 50 miles of Reading	2
Beyond 50 miles of Reading	1

If nothing else, this highlights the suitability of the location of Central Berkshire as a growth area: there appears to be little difficulty for the major builders to control operations from the nation's capital.

Block 4 The medium-sized housebuilder I

'I am a benevolent dictator.'

Here is the archetypal housebuilder with an air of overambition, ruthlessness and just a hint of underhand dealings. As the chairman of the board, our interviewee is the key decision maker and formed the company some twenty-one years ago. Interests then were in contracting but the succeeding years saw narrow diversification: property, plant and appliances. Housebuilding is of long standing (eighteen years) and is now the most profitable of the enterprises producing one-third of company profits.

Still with the feel of a family firm, the company continues to be loyal to a small nucleus of direct-employed local craftsmen. The full type and price range of dwellings are constructed on the full range of development site sizes. Recently output has been a consistent seventy dwellings per year but the property booms of the 1970s twice tempted expansion which ultimately threatened the firm's solvency. None the less, ambition is maintained and an annual completion rate of 200 dwellings is still sought. Diversification continues but the profitability of speculative housing despite its risks is too much of a temptation. But despite this diversification, funding is still something of a problem for speculative housebuilding. Other company interests produce insufficient profits to allow major transfers of capital within the company. With a housebuilding turnover of £2.5 million, outstanding loans (from a clearing bank) are likely to amount to £0.5 million.

The land search is his key to success. Most purchases result from personal approaches to landowners and the company is prepared to sit on a site for a decade before permission is viable. The land search is not locationally extensive but is highly intensive and undertaken through community contacts seven days a week.

Needless to say planning is the *bête noire*: 'I don't like planning officers but I can talk to them . . . they are bigoted little people.' The comments on the planning committee are unprintable. Attempts to fool them have included the submission of applications under pseudonyms.

Given a suitable economic climate, the aggression of this company may lead to expansion. But risks are entailed, sophistication of operations is not all that it might be, and this local building enterprise could succumb under the competitive weight of the major housebuilders who have been drawn into the growth area of Central Berkshire.

In contrast to the other sizes of companies, the operational areas of the large housebuilders are subject to sudden fluctuation. As a growth area, developers have been encouraged to move into Central Berkshire and the post-1974 date of entry of companies is estimated in Table 3.9. Notice how none of the large companies interviewed has moved into Central Berkshire since the official announcement of the modification of the structure plan requiring the land for the extra 8,000 houses (1980). Although this announcement caused a considerable public and political stir, housebuilders had recognized the area's potential for growth at an early stage: the request for the land for the 8,000 houses was partly due to their lobbying. The structure plan modification provided reinforcing rather than original justification for operating within the area.

TABLE 3.9: Entry to and withdrawal from Central Berkshire of the large companies according to the dates of their first residential planning applications

Entry to Central Berks	*Large companies*	*The five of the top ten*	*Withdrawal*
1974	2		
1975	2	1	
1976	2	1	
1977	5	2	
1978	2		
1979	3	1	1
1980			
1981			1
1982			

Only two large companies were known to have taken a decision to withdraw from the area during the study period. One withdrew because a policy of diversification of operational area had failed, while the other appears to have been heavily involved in early Central Berkshire land acquisition but for some reason, not convincingly explained, decided to withdraw.

Looking in more detail at five of the nation's top ten housebuilders, it is interesting to note that all bar one had arrived or reasserted their position in Central Berkshire by 1977. None, however, appeared to be involved in housebuilding in the area at the time of local government re-organization (1974). While the influence of most of these companies is increasing within the area, it is unlikely that any more of the leading builders will enter the area. Perhaps Central Berkshire has reached a saturation level of major housebuilders and the entry of others would unacceptably increase the level of competition. However, the final entrant was welcomed by some builders because of its strong marketing style – this, it was considered, could only help the industry in general and themselves in particular during a time of sluggish sales.

In summary, the interest of the large companies in Central Berkshire has risen markedly over the last decade. Most appear to have established their position prior to, or in the same year as, the publication of the Structure Plan Consultation Document in 1977 which clearly acknowledged the inevitability of further growth in the area. The aftermath of the allocation of the land for the 8,000 houses revealed no new entrants but it did raise hopes of increasing levels of business for those already in operation in Central Berkshire.

The labour component

Since the end of the last war, the total construction industry has undergone vast shifts in employment patterns. As part of the construction industry, the housebuilding industry seems to have undergone even greater shifts. The employees required by the housebuilder range from the administrative, secretarial, and sales staff to the large range of craftsmen and general manual site staff. The attitudes to each of these types of labour have varied considerably.

SITE LABOUR

The notorious fluctuations of the construction industry over seasonal and more long-term cycles have encouraged most housebuilders either to shed or never to engage directly employed site labour. The trend away from direct employment has continued or even increased during the current recession but as Leopold and Leonard (1983) have noted, this is perhaps less of a reflection of the recession than of long-term changes in the organization of employment in the construction industry.

The implications of the movement away from direct employment are considerable. Employers may pay more in the short term for a piece of work but this should be recouped in the long term since employers are effectively relinquished from many responsibilities and long-term financial commitment. A tied labour force is unwanted; to the larger enterprises the flexibility of being able to shed the workforce may be vital for survival whilst the very small builder may perceive it in terms of freedom: 'you can't be your own governor if you direct employ.'

Direct employment is most common in fairly large companies with interests in contracting and with the very small builders whose steady building rates can sustain a workforce. But even in the large companies, there is often a reluctance to use the direct labour of the contracting side of the business: such resource shifting may be tempting but often sets up tensions within firms (these tensions will be referred to again later).

Direct-employed labour is often used to undertake or monitor quality workmanship. The drift away from direct employment has had serious repercussions upon training and apprentice schemes and it has often been claimed that the quality of workmanship has suffered. Housebuilders (amongst others) find it difficult to ensure quality workmanship through contracted labour. Many firms have therefore employed site managers sometimes supported by finishing foremen (who act as monitors of quality control) on their often dispersed sites. Occasionally, one or two general labourers are employed for sundry site tasks. In trying to combat the same quality problem, some small firms direct employ for specific tasks to ensure a quality product. 'Groundwork' is a good example of this – groundwork marks the beginning of the job and although it is not a particularly skilled job, it must be undertaken with care to ensure a quality final product. Of the eleven businesses in which at least one of the administrative staff was involved in site work, ten specifically took a leading role in the groundwork. Other types of direct labour were used according to preference.

In the main, however, self-employed and subcontracted labour is the norm. Such labour is rarely viewed in terms of numbers but rather in terms of work done. A job of work is put out to tender and may be taken up by a major contractor or, more likely, many small craft-based subcontractors. It is somewhat ironic, however, that particular gangs frequently work in sustained contact with particular housebuilders. There is an inbuilt inertia in the system which leads not to massive labour switches but rather to a semi-stability. Occasionally it even goes to the extent of housebuilders gearing their operations around the availability of particular workers or gangs.

ADMINISTRATIVE STAFF

Arrangements in the employment of administrative staff are also changing. There has been a more recent drift away from direct employment towards the use of agencies – especially with regard to planning matters. It is more appropriate to discuss such internal organization of firms in a later section but here it is useful to indicate the number of non-manual site labour employed (Table 3.10). Attempts were made to include only those staff employed in housebuilding, but inevitably there were overlaps and the figures should only be taken as guides.

The most obvious point to be made is that speculative housebuilding has remarkably few direct-employed staff. Most of the very small companies are run by individual entrepreneurs and most of the small and medium-sized firms directly employed less than ten

TABLE 3.10: Number of staff (excluding site labour) employed by housebuilding companies

	No. of staff							
	1	2–4	5–9	10–24	25–49	50+	N/A	Total
Size of firm								
Very small	7	2	1					10
Small	–	4	5					9
Medium	–	1	2	4	3			10
Large	1	–	1	3	3	4	2	14

staff. Even some of the large companies managed with remarkably few staff and only the major builders had staffs of more than fifty.

The dwellings constructed

The range of dwellings constructed by the housebuilder can vary enormously. When design and layout were of lower priority, the classification of product was relatively simple, e.g. detached houses, detached bungalow, semi-detached houses, terraced houses, attached bungalows, flats and maisonettes. In fact, this classification is still currently used by the NHBC in their statistical quarterly but appears rather inappropriate in the context of Central Berkshire where linked detached are frequently found on the large estates and where neo-back-to-backs are beginning to make an appearance.

It has proved very difficult to quantify the type of output of housebuilders in the sample. Figures for the medium and large firms were seldom available. Dwelling type information proved unsuitable for analysis and here will be discussed only in broad terms. Dwelling price, however, has been indicated in general terms (Figure 3.2) but since the figures are restricted to Central Berkshire and for the years 1981/2, the perspective offered is limited.

First, then, it is worth providing a general picture of the types of dwellings constructed in Central Berkshire in 1981/2. But to begin in the negative: bungalows, flats, and maisonettes were seldom built. Bungalows occupy too much land in relation to their usable floor space and so in an area of growth where land is expensive and in short supply, single-storey dwellings were only built where they were virtually necessitated by planning authority considerations. In practice, there was often compromise between single- and double-storey dwellings by the production of 'chalet bungalows' (with restricted second storeys). Bungalows were rarely found on the major sites but tended to occupy single-

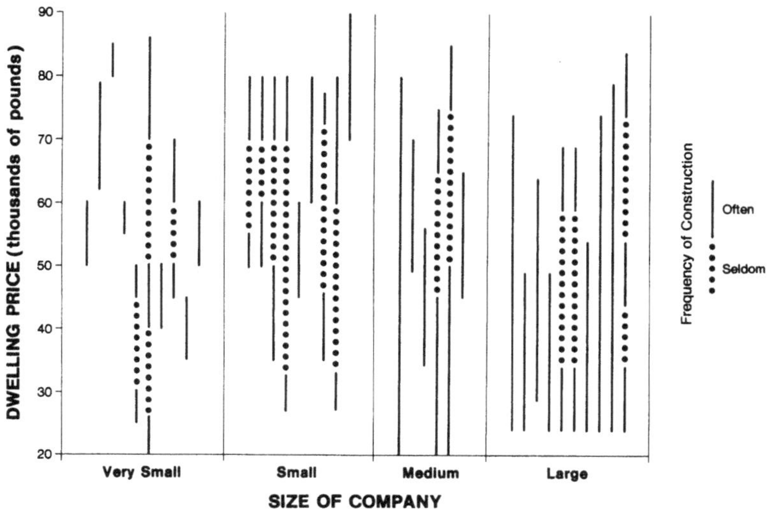

FIGURE 3.2 *The price of dwellings constructed by different sizes of housebuilder*

dwelling or very small sites. As a result, bungalow production was largely the concern of the small builders.

If land is so expensive and scarce, it might be expected that flats and maisonettes would be very popular but these formed only a very small proportion of builders' output in Central Berkshire; it was only the very small and small builders who probed this market. Sites for flats were only really available in the urbanized areas – in practice, Reading – but appropriate sites (in planning terms) were few.

Detached, semi-detached, linked, and terraced dwellings formed the vast bulk of production. Terraced developments were of course seldom advertised as such – 'link detached' were the order of the day.

In 1982, the second-hand terrace market price in Reading ranged from £20,000 to £25,000 and the most commonly encountered upper price was in the region of £90,000. It was easier to discriminate housebuilders' production by price than by type but the notion that particular builders exploit particular parts of the market needs some clarification.

As demonstrated in the discussion of the land search, housebuilders do not simply select a price range to work within. Through their purchase of a site, they are committed to produce a particular price range of dwellings. Planning authorities attempt to control the density of dwellings upon a site and density is highly correlated to price. In Central Berkshire in 1982, the first-time buyers' market (the cheapest range) was deemed to entail the quickest sales and therefore the most reliable and fastest profits. The higher the price the more sluggish the sales and therefore any benefit from a higher profit margin of the more expensive houses was effectively offset by a slower turnover with its accompanying capital-sapping interest rate payments. So the pressure within Central Berkshire has been to push for high density (and therefore relatively low-priced) developments. This can be admirably demonstrated by the willingness of the larger developers to take to appeal refused applications for high-density developments on the large sites. Increased densities are well worth obtaining.

From Figure 3.2, it is apparent that the very small builders tended to concentrate on quite narrow price range bands. As noted, the nature of the data would exaggerate such a notion since very few dwellings would be included. None the less, the finding is valid in restricted form. The most popular price range for the very small housebuilder was £40,000–£59,000. Anything more expensive than this risked exceedingly slow sales and few managed to break into the first-time buyers' range because of the competition for, and low availability of, suitable sites. It is often surmised that the small housebuilders concentrate on expensive 'luxury' sites (see, for example, Short, 1982b). The Central Berkshire survey offers no support for this. It is true that a few very small housebuilders do aim for the luxury market, but their numbers are few. But it is also true that a great many very small and small housebuilders would *like* to build for the upper reaches of the market. Only one very small housebuilder managed to span the full price range and it is no coincidence that this particular company was expanding and ambitious. With this exception, the rule of the conservative, restrained very small housebuilder is amply upheld.

The small housebuilders were only a little more likely to build through a greater price range of dwellings. There was a bimodal split in their output but few managed to exploit the first-time buyers' market or to span the small range.

The medium-sized companies fell into two groups: those who spanned the full range and those who concentrated on the middle bands. Through their operation on the large sites, three had managed to exploit the first-time buyers' market.

The large companies were quite different. Through their dominance on the large sites, they were enabled to span wide ranges, and to exploit the lowest end of the market where profit margins might be slim but fast. Their diversification of output, if not company policy, was virtually ensured by planning policies which attempt to provide 'mixed' developments. Such policies lead to a whole range of dwelling prices within small areas. However, it is interesting to note how some developers tackle the problem of providing mixed density developments: a proposal approved recently by Bracknell District Council produced a final density of twenty-seven dwellings per hectare by including 148 (40 per cent) four-bedroom detached dwellings and as many as 116 (31 per cent) two-bedroom back-to-back (quarter-detached) dwellings. Thus the developers had managed to include a large number of starter homes by also including a large number of up-market homes with larger than usual gardens but at the expense of minimizing the middle-range housing production.

Timber frame construction

Advances in UK housebuilding techniques have rarely been rapid or radical. The industry has traditionally been regarded as conservative and undercapitalized but over the last decade there has been a significant rise in a particular kind of industrialized production – timber frame. The attitudes and approaches of housebuilders and consumers alike provide an intriguing perspective of the industry.

Timber frame construction is not new. Many countries (with plentiful timber supplies) have witnessed this form of construction throughout this century. Indeed, even in Elizabethan England, timber frame was the major form of construction.

Timber frame construction at its most sophisticated is simply kit construction. Although the building is eventually clad in traditional brick, these materials are merely cosmetic – the timber frame is itself free-standing. The frame is manufactured off-site on traditional industrial premises by semi-skilled workers who need not be traditional construction sector workers. The frame is delivered to the site when required and can be erected in one or a few days. The dwelling is therefore made rapidly waterproof and the internal finishing need not be interrupted by the weather. Moreover, the housebuilder is freed from constraints of the availability of many construction tradesmen: no longer is the availability of bricklayers crucial to the timing of construction.

The advantages of timber frame centre around its speed, relative simplicity and the flexibility it affords to planning a development. The erection of a traditional brick and block dwelling normally entails a construction period of at least twenty weeks; timber frame dwellings can be completed within six weeks. Thus a housebuilder's capital need not be tied up for such a long time and, if required, the traditional adage of 'build and sell' may be reversed to one of 'sell and build': no construction need begin until a buyer is available and so a proportion of the risk element is lessened. Much tighter planning of a development is therefore possible. Lesser, but notable advantages include a greater sound and heat insulation of the dwelling, reduction in design error, weather-free construction, and reduced administration and supervision of the workforce.

The disadvantages are less easy to define. Timber frame units usually entail slightly higher costs and these costs are mainly borne at the beginning of the building process rather than being spread throughout the construction period. So although the duration of a loan can be kept fairly short, access to a substantial capital reserve is necessary. While this may not unduly affect the larger operators, it is often of concern to small firms. Moreover, timber framing is not particularly appropriate for sites of less than ten dwellings. The process is more subject to the volatility of the timber trade and uptake of timber frame is often related to these fluctuations.

Although mass-produced, timber frame dwellings cover the full range of house prices and do not have the stigma of low-priced, low-quality dwellings. However, consumer resistance has been high and grew considerably in the summer of 1983. The roots of the consumer resistance are difficult to trace but seem to originate from an inherent conservatism. Even though building societies quickly accepted their viability and the NHBC offered no resistance, consumers have not been entirely convinced. The industry has recently used a high profile approach in marketing timber frame and until June 1983 seemed to be gradually breaking down resistance. However, in June 1983 a 'World in Action' TV programme sent shudders through the housebuilding industry. The programme highlighted inadequate site work and poor safety standards associated with timber frame builders. The stock market reacted rapidly and unfavourably towards timber framing and a real threat to the method was perceived. The immediate media coverage was surprisingly slight but the issue built up and provoked major disagreements amongst the producers. There were allegations that the brick-and-block lobby promoted the anti-timber framing sentiments whilst the timber frame camp was itself fragmented and tensions were evident between the volume builders and those directly responsible for the manufacture of timber frames. The HBF was alarmed enough to take action to try to prevent inter-builder strife (*Building*, 19 August 1983, p. 11). Reports were commissioned but they only partly settled the dispute and its ramifications rumble on. The immediate effects have been to cause a lowering in the timber frame market share.

The uptake of timber frame In Scotland timber framing has accounted for at least 20 per cent of housing production since the mid-1960s; it currently runs at around 50 per cent. In England and Wales, uptake has been later but the market share had increased and in 1982 its share was 20 per cent.

There are marked regional disparities in the use of timber frame in England. Its market share is greatest in the South East and particularly in London where demand is greatest and the largest builders most active. No figures are available for Central Berkshire but it is probable that the proportion there is comparable with the rest of the South East.

Block 5 The medium-sized housebuilder II

'The industry is regimented like the army – but try to make us look human when you write about us.'

With a land bank dating back fifteen years, this local family firm was like the proverbial oil tanker heading for a sandbank. Constructed half a century before and with little new components since, this tanker had little chance of changing direction and seemed certain to be stranded without land. From a completion rate of sixty in the early 1970s, construction was gradually slowing and output had halved in the ensuing decade.

With many traditional values – tradition brick, avoid external borrowing, contracting and speculative construction do not mix, low-profile marketing – the company has never come to terms with post-1974 planning. Incredibly, the company lodged its first planning appeal on a purely residential item in 1983.

Local contacts abounded and much store was placed by personal interactions but these seldom appeared to be more than pleasantries. Hard and subtle negotiation was absent in all spheres of activity and the company had only just begun to twig the new professionalism of planning.

There is also a major disparity in the use of timber frame units by different sizes of builder. The national figures show quite clearly that the larger builders most favour timber frame: this fact is also reflected in the Central Berkshire sample. But not all the major builders use timber frame. There is a tendency to use it either for most production or only on a small trial basis – there is little compromise between and yet very few companies are entirely committed to the method.

What prevents a company from using timber frame construction? The conservatism of the industry is profound and this is particularly so amongst the small and medium-sized local companies. The site size favoured by the very small and small builders is not usually suited to kit construction. The absence of timber frame construction amongst the medium-sized firms is perhaps not so surprising in the light of other findings of this survey which indicated the frequent stagnation and relative lack of sophistication in this size of firm. And clearly, some of the large companies are waiting to see the prospects of timber frame. Those who are geared to growth have tended to adopt timber frame whilst the others are content with a steadier output.

Timber framing, then, is a contentious issue. It is an innovation which has been slow to take off in the UK despite the work of the timber frame lobby in the trade press. Every surveyed builder in Central Berkshire had at least considered timber framing but many were left unconvinced. While one company director of a small-sized firm noted, 'by the year 2000, 80 per cent of all new homes will be timber frame, I don't wish to be left behind,' others doubted its dominance. The industry's reaction to timber frame criticism has been edgy – it has been more concerned with the deteriorating public image of timber frame than its possible faults. As the researcher of the 'World in Action' programme commented, 'the industry is dominated to an unhealthy degree by unbending othodoxies and often it becomes far too sensitive to normal criticism' (*Building*, 22 July 1983, p. 23).

Housebuilding on large sites

A significant proportion of housebuilding in Central Berkshire is taking place on large sites (see Figure 3.3). Almost 16,400 houses will be built in the period 1977–1986 on seven sites, 6,000 of them on one site, Lower Earley. Large site developments have occurred elsewhere in Britain including Bowthorpe near Norwich, Goldsmith Park close to Woking and New Ash Green in Kent; for a discussion of this last development see Bray (1981). There are a number of forces at work. On the one hand there is an effective demand for new owner-occupier housing especially in the more affluent counties surrounding London. On the other hand there are economies of scale for the volume housebuilders in developing large sites where costs are minimized and profits maximized. At the local planning authority level, there are clear benefits in encouraging residential development on large sites. Development pressures are easier to handle and monitor while planning gain is more easily achieved. This gain is the ability of local planning authorities to pass infrastructure costs on to developers. The volume builders operating on large sites can more easily accommo-

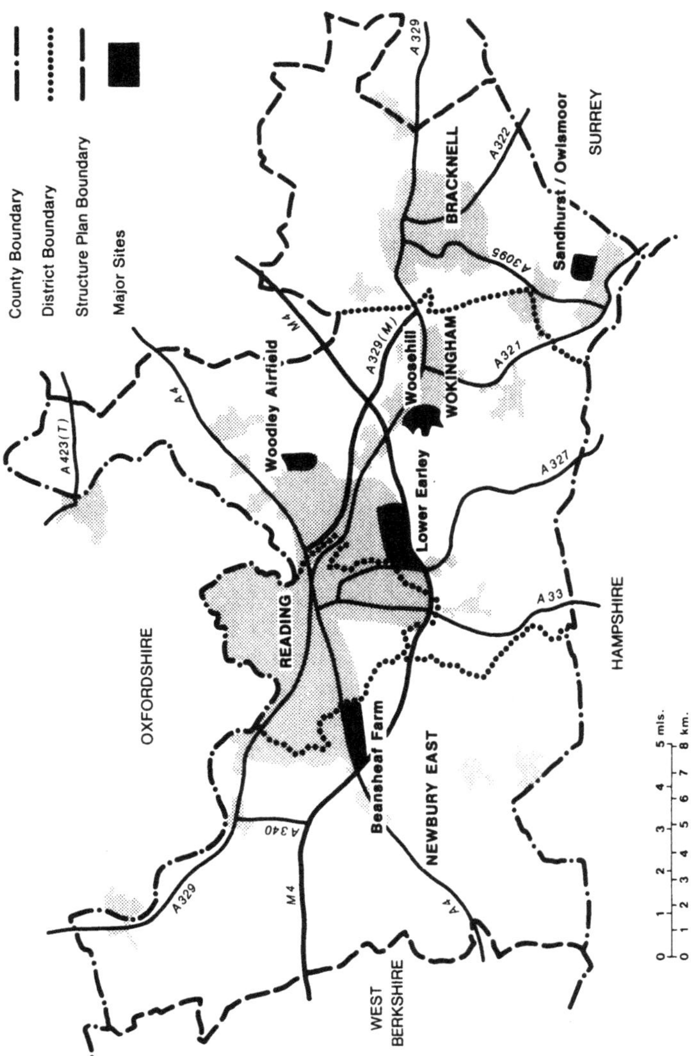

FIGURE 3.3 *Major residential sites in Central Berkshire*

date the costs of planning gain for any one site because of the volume of units they produce.

Lower Earley

Although all these large site schemes have their own particular story, they do share a number of general characteristics. As an example we can consider the case of Lower Earley just outside Reading. This site is within 'Area 8', one of five subregions identified in the 1970 strategic plan for the South East as an area of growth. Development plans were pre-empted by the Secretary of State's decision in 1969 to release land for residential development. At this stage sites could have been compulsorily purchased and development could have been undertaken by the local authorities with either all public sector housing or a public-private mix. However, this would have been against both national trends and the local political climate. In both the county and the district, Conservatives were dominant and emphasis was placed on the private sector.

Plans approved by the county and district councils in 1972 envisaged a consortium of landowners/developers coordinating development on a 1,000 acre site between the M4 and the southern boundary of Reading. After alternative reports had been published by a group of landowners and further negotiations were held, a single planning application was submitted in 1974 by a consortium including the University of Reading, Bovis Homes Limited, North British New Homes Limited, Madrey Properties Limited and J. A. Pye Limited, ultimately to involve the construction of 6,000 private sector dwellings.

Lower Earley marked a negotiated compromise between the local planning authorities and landowners/developers. Negotiations centred on the planning agreement, the development brief and discussions over subsequent planning permission. Under a planning agreement signed in July 1977, the landowners/developers agreed to pay for some of the road construction costs, provide land for other roads free of charge, to phase the developments in four stages, to provide public open space land free of charge and to make a minimum cash contribution of £660,000 towards infrastructure costs, the final amount being related to a percentage of house sales prices (see Henry, 1982). The agreement signals a substantial degree of planning gain achieved by the local authority. The development brief was produced by the local planning authority mainly for the developers, to show how development should be carried out. It was not a statutory device but more a series of advice notes relating to the mix of houses, the design of neighbourhoods, the amount of public open space and the size and extent of public facilities. An assessment of the brief by Thompson (1981) shows that it was only really effective in physical planning terms. The brief for Lower Earley was an attempt to move away from the endless sprawl image of private suburbs towards a concentration of small (down to five houses in some cases) groups in cul-de-sacs, in an attempt to produce small-scale environments which it was hoped would foster community ties. The housebuilders' favourable response to this element of the brief was mainly due to the confluence of interests. For them the small, cul-de-sac developments proved easier to sell than the individual houses in a monotonous row of anonymous streets.

With respect to socio-cultural facilities such as community centres and public services, the brief, lacking any legislative purchase, was more a statement of hope than an effective blueprint. The result was for lots of houses and few facilities, and some local press response to Lower Earley is reminiscent of the early publicity received by the new towns. An article in the *Reading Evening Post*, for example, in September 1982 was headed 'Housing estate is becoming a nowhere place'.

Since 1977 there have been subsequent planning applications on the Lower Earley site. At the pre-application discussion stages the planners have sought to enforce the brief. The main point of dispute has centred on density levels as housebuilders seek to increase densities, which generates greater profits. Planners have sought to resist this trend.

In an advertising feature in the *Reading Chronicle* (30 October 1981) under the general title of 'Move to a new way of life in an Abbey home' the following points were made.

> the design of the site was based on special design guide principles which set out not only to create a village style environment for the development, but also to minimise major traffic flows. Small clusters of homes are planned with short cul-de-sacs and private drives. Another pleasing aspect of the development will be the use of different treatments and materials for the elevations of the many different designs of houses. . . . It all adds up to variety and a pleasant 'different' and villagey-look.

The Lower Earley site is subdivided with units between ten and thirty houses. The major national and regional builders including Bryant, Abbey, Pye, Laing, Costain, McLean, Wimpey, Gough Cooper and Bovis all have units. Those units differ in building material, layout, price and external design. A typical cheaper-priced unit consists of thirteen three-bedroom semi-detached houses costing (in August 1981) £34,650. An average medium- to high-priced unit consists of twenty dwellings, three semi-detached for £47,250 and seventeen detached ranging from £54,000 to £67,500. Over 80 per cent of the dwellings have either three or four bedrooms.

Given the size and prices of dwellings in Lower Earley, the social composition skewed towards higher income groups. Surveys of residents have shown that just over 50 per cent of the levels of households come in the high income professional and managerial occupational categories and less than 5 per cent are in the unskilled manual category. Lower Earley is a residential area catering for the middle and upper income groups.

The marketing and selling of new housing

The interviews with the housebuilders in Central Berkshire were undertaken in a period of deep recession, high but declining interest rates and very slow house sales. These conditions created unusual marketing problems and elicited novel responses. The replies to the question on marketing may therefore be untypical although there is reason to believe that the 1981/2 slump in house sales may have altered the marketing of dwellings for years to come.

The image of the new house

A new dwelling is likely to be the single most expensive purchase of most UK households. Given this fact it is surprising to note the relative lack of positive quality control. Only recently have there been serious suggestions that there might be room for a positive quality control scheme such as is operated for other consumer durables by the British Design Standards Authority.

The industry has not encouraged quality design. And much of the medium- and low-priced private sector housing built between the mid-1950s and the mid-1970s was of poor quality.

However, there was a major improvement in housing design and layout in the 1970s. The Essex Design Guide of 1973 which laid down guidelines concerning external design and residential layouts has been eagerly adopted by many local authorities. Gone are the regular rows of identical housing. Here (and, some would argue, with equal monotony) are the more complex layouts of cul-de-sacs arranged in a dendritic-like pattern. A greater degree of dwelling mix has been encouraged and developers now appear to find it advantageous to advertise their dwellings by maximizing even the smallest differences between houses.

Unlike most consumer durables, the new home has found its virginity to be a handicap. Dwellings and layouts are perceived to mature and to prove themselves with age and the industry has had to fight against the images of cowboy construction and building-site living. The industry has therefore found it necessary to promote the newness of its product although few individual companies have explicitly recognized through their advertising that the second-hand homes market offers serious competition.

To combat this, the HBF launched the New Homes Marketing Board (NHMB) in 1982. The NHMB is attempting to raise the profile of the new home for two reasons. First, and most directly, it is trying to improve the image of new home to make it attractive to the consumer. Second, and though indirectly, of great importance, it is attempting to make the new home acceptable to the existing residents of an area. The HBF has recognized through its dealings in the land availability debate that one way to alleviate that problem might be to present the attractive side of new housing development. In this way it hopes to win the 'hearts and minds' of the anti-development lobby.

The image of the residential area

In recent years the housebuilders, and particularly the volume builders, have also been concerned with selling individual dwellings as part of a residential area. The unit construction, noted in Lower Earley with its cul-de-sacs and dwelling clusters has both forced and enabled builders to sell the immediate area as well as the individual dwelling. The main effort has been directed towards tapping popular images of rural harmony and village cohesion. This is evident in the naming of residential units in Lower Earley. As Table 3.11 shows the housebuilders have employed an explicitly rustic vocabulary. The names themselves are loudly emblazoned on large billboards with pictures of swooping swallows and giant daffodils or placed above red-tiled wells which tap no known water supply. On the large sites such as Lower Earley there is a conscious effort by the housebuilders to 'villagify' the residential areas. The attempt is to draw upon the rich and long-established English rural mythology in order to sell new houses in brand new residential areas.

The incentives to buy

The early 1980s and the initiative of a single major housebuilder has radically altered the marketing and advertising of the new dwelling. More than any other feature, the Barratt helicopter has

TABLE 3.11: Neighbourhood names in Lower Earley

Hunters Walk	Ryhill Copse	Upper Paddock
Badgers Walk	Abbey Lea	Pipers Dell
Greenbanks	Mead Ridge	Swallows Meadows
Millers Green	Meadow Vale	Hedgefield
Upper Wood	Meadowbank	Spring Fields

signalled the change in the industry's marketing of its product. Gone are the days when the new home was 'adventurously' marketed as a barren shell – in an attempt to boost sales on the high-quality SPAN development of New Ash Green, undecorated units were offered to first-time buyers at a lower cost (Bray, 1981). Today, fully decorated and equipped dwellings are offered with all manner of incentives and inducements.

The sample of available incentives shown in Table 3.12 indicates the anxiety of the industry to sell its product, and the market at which they are aiming. The incentives package has been the response of the large housebuilders to the slow sales of the recession. Only slowly and to a very small extent have the smaller housebuilders been drawn in to offer similar deals. Generally, those who were drawn in were the most ambitious and sophisticated. But most of the smaller firms were reluctant to offer 'gimmicks'. A director of a local medium-sized company stated his case: 'We have purposely stayed off these types of incentives . . . we argue that the house is value for money . . . basically, we will not negotiate with the consumer . . . the houses sell themselves.' But do the large housebuilders consider that the incentives increase sales? As we have seen, the incentives are principally aimed at the first-time buyer and are geared to speeding up sales. At the end of the day the cost feeds back into the price of the house. (House valuation problems were evident in the summer of 1983 when consumers began to realize that on the resale of new homes it was not always possible to recoup the original selling price which had included the costs of incentives and internal fitments.) The investment potential of the house was therefore threatened. One large company which claimed to have introduced the incentive-type schemes in the early 1970s had abandoned them claiming that they were expensive and ineffective. Others admitted that they had been pressurized into offering incentives packages for fear of being left behind by competition. Only a few considered that they were positively helpful in increasing the number of transactions – most seemed to tag along for fear of losing a relative share of the market. It was quite clear, however that little market research had been undertaken.

TABLE 3.12: Sample of incentives offered by housebuilders in Central Berkshire in 1982

Reduced or no payments necessary for solicitor's fees, mortgage survey, and/or stamp duty – often amounting to £250–500 saving on £20–30,000 dwellings. In practice these 'savings' are tacked on to the price of the house, and the consumer pays via the dwelling mortgage payments.
100 per cent mortgages – no deposit necessary – thus aiming at the first-time buyer who has little or no capital. Sales were so slow and interest rates high in 1982 that the building societies (often through their quota arrangements with the housebuilders) could offer attractive mortgage arrangements without fear of being swamped by demand.
Mortgage repayment protection schemes – an elaborate insurance scheme if the consumer loses his/her job.
Fully or partially furnished homes usually with many white electric goods. But again such 'extras' merely end up tagged on to the house price and mortgage.
Payment of lump sums to the consumer, often into building society accounts on completion.
Special discounts for quick exchange.
Reduced mortgage rates for limited periods. But this merely spreads the high interest rate costs of the early years into later years.
Part-exchange schemes vary considerably but often the new home builder will offer to buy the purchaser's old home at 95 per cent of its 'independently' valued price. This can often involve the new housebuilder in the sluggish sale of old dwellings but since the consumer is 'stepping up', the sale of the old dwelling if in a suitable area should be easier. In any case, stamp duty is only payable on the difference in price between the two dwellings and the housebuilder is buying at only 95 per cent of the independent valuation.

The point of sale and the use of estate agents

The use of estate agents in finding land for housebuilders has already been noted (see the section on the land search, p. 50). In many cases 'gentleman's agreements' are usually struck which give the agents the opportunity to sell the dwellings built upon land which they have brought to the attention of the housebuilder. But estate agents are also often called upon to sell homes in circumstances where they had no involvement in the acquisition of the land. In fact, most housebuilders use estate agents for nearly all their sales and it is really only the large companies which organize themselves in such a way as to incorporate the sale of houses into their own activities. With the increase in marketing awareness, however, the direct sell is becoming more common.

The ability of estate agents to sell houses has increasingly come under fire (see, for example, the July 1983 issue of the *Housebuilder*). Notably, estate agents do not have the status of a profession and many of the smaller housebuilders interviewed articulated a certain disdain of estate agent's activities – basically agents were accused of not trying hard enough.

But, of course, estate agents are variable and some were recognized as good land finders and others as good home sellers. No single estate agency operates on a national basis. None exclusively deal in new dwellings but one particular agency in the Reading area had a specialized new homes department and was used by both local and major developers. A few of the major agencies have established arrangements with certain industrial companies who are in the process of relocating to Central Berkshire. Through such arrangements company moves are facilitated and housebuilders have relatively easy access to the incoming consumers.

Despite these type of advantages, many of the major housebuilders prefer to control their own sales. Although the necessary sales staff and office will entail overheads (but the sales office often doubles as a showhouse), direct liaison with the customer gives the company a better understanding of consumer preferences, allows greater marketing control, and avoids the estate agency's commission fee (which varies by negotiation from 1 to 4 per cent). Only on ten-dwelling sites or smaller will most of the larger companies contemplate the use of an agency.

It might therefore be expected that serious tensions would exist between the housebuilders and the estate agencies. The agencies' involvement at both ends of the development process would appear to give the agencies a degree of leverage power. Such power is probably exerted with effect on the small housebuilder – any builder who refuses to comply with the gentleman's agreement or who builds slowly or constructs dwellings with marketing problems can be hindered by being deprived of the estate agent's supply line of land. The extent to which this occurs is difficult to monitor but it undoubtedly occurs. Nearly all of the small housebuilders did remark upon the influence which the agencies did try to exert through their 'underpricing' of dwellings.

However, the agencies' attitudes to the larger housebuilders appeared much more deferential. Despite the fact that the agencies seldom have the opportunity of selling the major developers' houses, they try to maintain good and informative relationships because it is recognized that today's new dwellings are tomorrow's second-hand sales. And second-hand sales form the bulk of an estate agent's trade.

Provision of the customer's capital

Although the Abbey National Building Society in 1982 initiated moves to build their own houses, there has been no counteracting tendency for housebuilders to provide consumers with long-term loans. The industry needs its capital for its own activity.

However, many companies have agreements with building societies and/or banks, whereby the housebuilders receive quotas to supply their customers with mortgages (Table 3.13). Such agreements are confined almost exclusively to the medium and especially the large companies. No fees are necessary – the lenders need to lend and the sellers need to sell – each benefits from its own operations and the agreement is mutually acceptable.

TABLE 3.13: Companies with building society or bank mortgage quota agreements

	Building society or bank quota agreements	*No quota agreements*
Very small	1	9
Small	0	10
Medium	4	3
Large	12	1
	17	23

The influence of the building societies (and now the banks) upon the activities of the housebuilder is not readily apparent. But influence is exerted. The societies require sound long-term investments and lend only on certain types of property. Since few purchasers can avoid borrowing, the orthodoxies of building materials and methods imposed by the societies are profound. Few housebuilders recognize the societies' influence – that fact emphasizes the extent of their control. One of the more adventurous housebuilding ventures of the 1960s and 1970s was SPAN's 'village' development of New Ash Green. One of the reasons for its failure was the reluctance of the building societies to offer mortgages on their radical materials and design. More recently, it has been intriguing to note that the building societies stood firm in their support of timber frame when that building method came in for some heavily publicized criticism in July 1983. Had the building societies rejected timber frame, the operations of even the biggest housing developers would have dramatically altered. But the mutual interdependence of the societies and the major builders acts to prevent major differences in policy.

The housebuilders' responsiveness to their consumers

For all the expense involved in the purchase of the new home and the recent rise in awareness of marketing, the 'absent consumer' has remarkably little influence upon the nature of the product. The consumer choice is largely limited to the acceptance or rejection of the product. Obviously housebuilders get some feedback about their product but only one major housebuilder is known to undertake a systematic survey of all their consumers' views of their new purchase. The general lack of market awareness is profound. True, today's estate layouts are much more adventurous than even a decade ago but initiatives have come from the top down.

Few homebuilders countenance other than cosmetic changes and amongst the small builders it is clear that extraordinarily few houses are custom-built: described as 'one of the most original ideas to hit the housebuilding industry for many years', one company has launched the 'flexi-home', a construction which offers the consumer up to fifty permutations of internal layout.

Generally, the industry is reluctant to respond to the consumer:

> 'We refuse to alter the planned design and we eliminate all choice where possible . . . we don't offer choices, the houses sell themselves.'

> 'Five years ago I had a Saturday morning job of finding out the sorts of alterations individual customers wanted. But there was too much aggro and no money in it. Today, we treat them like children in school – you don't want to start giving them choices!'

A few even hold back sales until the plastering is completed and alterations are impossible, despite the security of the early sale.

In the course of interviewing, very few interviewees volunteered or stressed the quality of their product. Pride in the job was infrequently expressed. There were of course exceptions but these served to strengthen the view that the industry with its heavy reliance upon subcontracting and piecework does not place much emphasis upon quality to sell its product. One major public company has forged itself a place in the market through its promotion of quality, but one very small builder lamented, 'few purchasers know much about houses and simply don't appreciate quality workmanship.'

The introspection forced upon the industry by the recession has led to a recognition of the problem of less than adequate design and workmanship and according to the Housing Research Foundation only 43 per cent of the housebuyers would recommend their builder to a friend. The NHBC, in an attempt to push the positive side of quality control, has launched a 'Pride in the Job' campaign. As noted, the New Homes Marketing Board has been established, and the Housing Research Foundation has stepped up its market surveys. But the industry is having problems in shrugging off the dual images of land speculation and cowboy construction.

Capital provision

Although some may use internal funding, most speculative housebuilders are not able to operate without some form of external source of finance. The price of land (in 1982, often as much as £35,000 for a single plot or £60,000 for one hectare on the large sites) necessitates the tying up of capital for substantial periods of time prior to realization.

Self-financing

In the total sample, only three firms claimed not to use loans (see Table 3.14). Two of these were very small: one could truly be classed as a 'stagnant entrepreneur', while the other was gradually withdrawing from speculative housebuilding. On occasions the pride of the small entrepreneur was apparent and some viewed independence from borrowing as a highly desirable but unattainable goal. Yet such independence militates against increased output and growth, and it appears not to produce higher levels of profit. Notably, neither of these very small firms was a registered limited company and they were the least formal and sophisticated of all interviewees. To a large degree, they were overgrown contractors with limited ambition.

The one other company which claimed not to borrow money was at quite the other extreme. A highly successful major housebuilder, this company did have external financing arrangements but claimed to have met all its needs through internal (speculative housebuilding sector) funding for the past two years.

As will become clearer below, it was difficult to determine the exact nature of internal funding arrangements and the extent to which parent companies might themselves by relying upon external borrowing. None the less, the low number of self-financing businesses in Central Berkshire does contrast sharply

TABLE 3.14: Sources of funding for housebuilding

	Self-financing	*Clearing banks*	*Merchant/clearing banks*	*Merchant banks*	*Internal funding*	*N/A*
Very small	2	8	–	–	–	–
Small	–	7	–	2	–	–
Medium	–	4	1	1	–	1
Large	1	1	2	–	5	4

with a study in Nottingham (Nichols *et al.*, 1981) where 23 per cent of companies were reported to be self-financing in the long term. This may to some extent reflect a marked difference in approaches to housebuilding in an inner city and in a growth area; the latter requiring a greater extension of credit, and attracting the more ambitious companies.

The clearing banks

For most housebuilders, clearing banks were the major source of loans. The theme of the builder-developer as a negotiator was particularly evident in this sphere and arrangements which existed were many and varied. The use of clearing banks was almost exclusively limited to the 'Big Four' (Barclays, Lloyds, Midland and National Westminster). While arrangements may have varied according to durations of loans and securities, astute borrowers often applied pressure to obtain more beneficial terms by emphasizing their past record.

One aspect of the loan which was open to negotiation was the gearing ratio for development costs and building costs. On development costs, two-thirds was often quoted as the upper limit of a clearing bank's loan while for the building costs, the figure was around one-third. But these limits were negotiable – so long as the applicant knew they were negotiable: one very small but long-established housebuilder was unaware that the 'Big Four' even considered loans on development costs.

Interest rates to be paid on loans varied enormously: the maximum encountered was 4 per cent over base, the minimum 1 per cent over base. The fixing of rates was usually related to the duration and security of the loan but the negotiating ability of the applicant was not inconsequential. Against the expectations of its accountant, one small firm had succeeded in cutting its interest rate payment to 2.5 per cent. In time this firm would press for a further reduction.

The policies of the clearing banks are flexible and differences in implementation by branch managers often caused transfers of allegiance amongst borrowers. The role of personal relationships was again an important element.

The influence of the clearing banks in the activities of the small builder is greater than is usually realized. By granting or denying loans, the banks act as 'gatekeepers' in selecting which companies will be facilitated in their entry into speculative housebuilding and which will be helped to expand. Even during the course of construction, the banks may attempt to retain a measure of control over what is, in effect, its investment. Site inspections are not uncommon. This control is highly variable but, in subtle ways, pressures may be exerted on slow or 'deviant' housebuilders.

The extent to which borrowers utilize the full facilities of their loans varied. Many small builders took a pride in underutilizing their loans. Loans varied dramatically by season and it was rarely possible for builders to quote an average loan commitment.

The merchant banks

The merchant banks were also utilized but to a much lesser extent. Although they generally enforced higher interest rates (often 4 per cent above base rate) than the clearing banks, it was reported that they were much more flexible and more willing to lend the large sums of capital necessary for large-site development. The roles of the clearing banks and the merchant banks therefore tend to differ and this was well exemplified by one medium-sized company which used a clearing bank for small infilling site developments and a merchant bank for its involvement on a large site. Whereas the influence of the clearing banks was felt mostly amongst the very small companies, the merchant banks were most closely involved with the larger enterprises.

Internal financing

Reference has already been made to self-financing companies and the difficulty in determining the nature of the arrangements in the large companies. Only five of the (large) public companies claimed that the capital necessary for speculative housebuilding came from their company group. These internal arrangements were seldom comprehended by the interviewees.

With the exception of the public companies, it is quite clear that speculative housebuilding, in Central Berkshire at least, is heavily funded through commercial borrowing. For many this is a necessity and for a few a preference. This reliance on external sources means that fluctuations in interest rates have a major impact on the industry. The precise effects – increased house prices, reduced profits or bankruptcies – depend on the state of the market and the way individual firms cope with these extensive shocks.

Block 6 A large housebuilder

'As an industry we must work on facts and figures – this is our only real strength – otherwise we have no political welly.'

As a very large national firm, this company has had a long history in speculative house construction. However, it has only been since the mid–late 1970s that their speculative housebuilding has expanded into the Thames Valley area. The company is ambitious but curiously is very dynamic in certain directions and rather conservative in others.

The dynamism of the company centres around its land finding and its emphasis upon political lobbying. Land finding is regarded as highly important and for the South East region four land negotiators are employed. Much effort is directed at contacting private landowners in attempts to establish option agreements. But there is no real effort to build up owned land banks – instead attempts are made to make large sites available through option agreements. This necessitates links with other housebuilders so that option land surplus to requirements may be disposed of. The company is a leading member of one of the two major development consortiums operating in Central Berkshire. It is also very active in the regional and national meetings of the HBF and sees the value of the political promotion of speculative housebuilding. In its dealings generally, the company has shown an unusual willingness to cooperate with other industries and developers and in this sense may be regarded as very progressive.

Against this progressiveness, however, there is a certain conservatism – resulting perhaps from being unable to direct its dynamism on all fronts. Timber frame has been tried tentatively but there is a strong adherence to traditional brick despite the fact that considerable efforts have to be made to link sales to completion rates. Borrowing is limited and it is claimed that the speculative housebuilding is self-supporting.

The importance of marketing is recognized but the attitudes towards estate agents are curious. All dwellings are sold through estate agents, yet the company takes a very strong directive role in the operation of the agents and even goes to the extent of training some estate agents in the art of sales techniques. It will be interesting to monitor how long this curious internalization and externalization of marketing will persist. The paradoxical attitude to marketing is perhaps generally indicative of the company's dynamism and conservatism.

The external contacts of housebuilders

So far we have been considering the internal organization of firms. However, it is equally important to note that firms interact with each other and with other agents and institutions. In the remainder of this chapter we will consider the institutional nature of housebuilder representation in the wider scene before examining the interaction of housebuilders with the land use planning system.

Housebuilders' organizations

No single organization has a monopoly position in representing the construction industry as a whole or even speculative housebuilders. Over the last decade however and particularly, over the last five years, one organization – the Housebuilders' Federation (HBF) – has emerged as the most important voice of private housebuilders. Its competitors are few and tend to represent the extreme ends of the housebuilding size spectrum.

The Federation of Master Builders (FMB) was founded in 1943 and has a membership of around 22,000. It tends to represent very small builders, some of whom are speculative housebuilders. As Rydin (1983) has noted, there has been some friction between the FMB and the much larger National Federation of Building Trades Employees (NFBTE) who have wished to include the FMB within their umbrella organization. Only a few of the very small housebuilders within the Central Berkshire sample were members of the FMB and appear to have first joined the organization as tradesmen and contractors.

At the other end of the scale, the Volume House-Builders Study Group (VHBSG) initiated by Tom Baron in 1975 consists of 'the chief housing executives of arguably the nine largest housing developers in the country' (*Building Magazine*, April 1982): Barratt Developments, Bovis Homes, Broseley Estates, Christian Salvesen, Comben Group, Leech, McLean and Sons, New Ideal Homes, and Wimpey. This group does exclude some of the major housebuilders but between them they account for approximately 35 per cent of all new homes for sale in the UK. Initially, many of this group were dissatisfied with the established organizations of the NFBTE and the HBF and considered that their interests could not adequately be represented by contractors or smaller housebuilders. Led by Tom Baron, who was a personal housing advisor to Michael Heseltine in his term as Secretary of State for the Environment, this group has been extremely influential in the DOE (see Rydin, 1983, for a detailed treatment of this influence). (Moreover, as will become clearer below, they have been extremely influential in mobilizing the HBF grouping within the NFBTE.) For some time, two of the VHBSG members (Broseley and Barratt) remained aloof from the HBF but they have recently joined and the future of the VHBSG in its current form may be in doubt. Six of the VHBSG members currently operate in Central Berkshire.

The Group of Eight comprises a variety of organizations concerned with the construction industry: NFBTE, Federation of Civil Engineering Contractors, Union of Construction and Allied Trades Technicians, Transport and General Workers' Union, Royal Institute of British Architects, Royal Institute of Chartered Surveyors and the National Federation of Building Materials Producers, Institute of Civil Engineers. Such a wide-ranging grouping is not always entirely relevant to the speculative housebuilders but their lobbying can sometimes be effective.

Beyond this, there are other groupings such as the Confederation of British Industries (CBI) but their interests are so diverse that their pressure is seldom directly relevant to the construction or housebuilding industry.

The National Federation of Building Trades Employers (NFBTE) was founded in 1878 and has approximately 12,000 members. As an organization it provides some very specific services for its members and also takes on a lobbying role. It has links with many organizations and its primary thrust is in representing the contracting side of the industry. It acts as an umbrella organization and the HBF has emerged from beneath this umbrella.

The Housebuilders' Federation

The Housebuilders' Federation (HBF) was established in 1939. Membership is open to all those who are members of NFBTE and whose names are registered by the National Housebuilding Council. (NHBC). Their relationship with the NFBTE is complex and has been subject to the contracting builder and speculative builder tensions which are present throughout the industry even down to the level of individual companies. Membership fees, calculated on a formula involving numbers of employees and annual turnover, are paid into a common NFBTE purse. There is no separate accounting and the costs of operating the HBF have never been calculated.

The role of the HBF has changed considerably over the years. In 1970 and 1973, the constitution was altered and its name changed from the Federation of Registered Housebuilders. It has always provided specific services to its members but it was not until the mid-1970s that the HBF began to take its public relations and lobbying role more seriously. A perusal of the *Times* index shows a sudden change in the type of entry with which it was associated. Prior to 1978, HBF entries usually concerned technical reports and fairly low-key statements. After this date, however, the HBF's role as pressure group became increasingly obvious with such headlines as, 'HBF president acuses local authorities of "intolerable efficiency" over planning applications' (*The Times*, 3 December 1981).

What caused this change? The HBF have claimed that in the late 1970s structure plans were not seen to be working and that there was a groundswell of member opinion which called for a stronger line to be presented by their organization. Notably, it was after 1975 that the exclusive VHBSG convened: they were clearly successful in their presentation of the large private housebuilders' case to the DOE. This must have made a particularly strong impression upon large housebuilders who were not part of the VHBSG and upon the smaller housebuilders who may have felt that their particular problems were being ignored and in fact exacerbated by the aid given to the volume builders.

Rapid change occurred within the HBF between 1976 and 1978. Roger Humber, a political scientist by training and a former research analyst for the Conservative party, took over the post of Director of the HBF. His forceful style became immediately apparent in the HBF's journal, *The Housebuilder*. The question of land supply was brought forward as the key issue where it has remained. Although *The Housebuilder* is supplied to every company within the NFBTE which is also registered by the NHBC, it reaches a wider audience by being distributed to others with interest or influence in the housebuilding industry. Far from being an in-house journal, it has formed one of the public relations platforms of the HBF.

Since 1978, the staff of the HBF has been increasing. The appointments have principally concerned Land and Planning Officers for each of the ten regions but in 1982 a full-time economist was appointed to give a broader balance of expertise.

The HBF has several functions. Its overall aim is to 'promote the interests and improve the status of private housebuilders' (*HBF Annual*, 1981, p. 5). It provides general information and detailed advice services for its members, but, its emphasis in the early 1980s was upon broader public relations activities and lobbying of specific bodies. It is these aspects which will be discussed below.

HBF officers have close and regular contact with the civil servants of the DOE. These links have blossomed in the period 1979 to 1983 and while Tom Baron's influence on the content of Circular 22/80 is widely acknowledged, the input of the HBF officers to Circular 9/80 is less publicized. HBF efforts concentrated on the issue of land supply and while it believed that it has won over the DOE officials on the technical issues, it was unable to persuade the ministers to take appropriate political action.

Several meetings are held with ministers each year at regular intervals at which the HBF case may be presented. At such meetings the information flow is dominantly one-way and the HBF can assess the value of the previous meeting by the ministers' initial statement of progress so far.

Realizing that the political battle will be long and hard, the HBF is currently encouraging and briefing its members on methods of lobbying their local MPs. A Parliamentary Liaison Team has been created and individual MPs have been mailed with information about the county land studies. Certain MPs are selected for particular attention and, at election times, those in charge of manifesto production are key targets.

At the levels of the county and district, the HBF focus is upon the chief officers rather than the elected members. With some planning departments, there is regular contact but due to historical accident there is no regular established contact with Berkshire CC. However, through the request of Circular 9/80 contact has been forged in the completion of the land supply studies. While aware of the local political personalities, contact with the members has been little pursued and a meeting with the members of Wokingham DC Planning Committee about the Lower Earley housing scheme was an isolated occurrence.

The HBF has had virtually no contact with local communities but they have become actuely aware of the importance of 'winning the hearts and minds'. To this end, the New Homes Marketing Board which has recently been set up is not narrowly aimed at consumers but is aimed at the populace as a whole in the hope of raising the profile and the acceptability of new housing. By so doing it is shaped to counter the 'anti-growth lobby' which has been increasing in strength over the last decade.

Links with the planning system

From a relatively low level of contact required by the Town and Country Planning Act of 1947, links between developers and the agents of the planning system have become elaborate, extensive and sophisticated. The original framework which aimed at regulatory control of development now appears to have a quaint naivety. The original lines of contact have become important two-way communication channels and other communication networks have been developed usually to facilitate applicants in achieving their ends.

In this section the nature of the links between housebuilders and their agents and the planning system will be considered. Definition of 'the planning system' will be broad and include those groups and individuals who register their views about planning matters. The different levels of the system range from the Department of Environment through county district and parish councils to specific groups and individuals.

A housebuilder's contact with the planning system may be at several levels (see Table 3.15). Since every development requires planning permission, all housebuilders must have some contact with the officers of the appropriate district authority. Because highways are a county matter and need to be considered for all residential applications, there must be some (though often very limited) contact between developer and the county officers (the county surveyor). In theory, and in practice for a few developers, direct contact with the planning system can be confined to written and verbal communications with the officers of the district and county. However, for the more sophisticated, ambitious or petitioned developer, contact with the planning system can occur at any of the levels indicated in Table 3.15.

TABLE 3.15: Possible points of contact between housebuilders and the planning system

	Appointed officers	*Elected representatives*	*Other*
DOE	R	R	
County Planning Authorities	F	R	
District planning authorities	F	O	
Parish authorities	R	O	
Groups and individuals			R

R = rare O = occasional F = frequent

Links with central government and the Department of the Environment

While the vast majority of planning decisions are issued by local authorities, the power wielded by central government is profound. Central government authority is expressed at the local level in four main ways.

First, the DOE issues circulars at irregular intervals. Circulars, advising local planning authorities on appropriate procedures and actions, are used by central government to coordinate national planning policies and frequently reflect the political philosophy of a particular administration. Although only advisory in nature, circulars carry considerable weight because their enforcement may be undertaken through the appeal mechanism (see below).

Second, structure plans and some other local plans devised by local authorities are subject to the approval of the DOE. The DOE frequently imposes modifications and again these may be backed up by the appeal mechanism.

Third, and more specifically, the DOE may intercept the development control process by 'calling in' major and contentious planning applications *before* the local authority has issued its decision. The 'call-in' procedure, however, is seldom used.

Fourth, the appeal invokes central government authority. An applicant may initiate an appeal to the DOE against a local authority development control decision. Appeals are frequent (there were almost 14,000 submitted in England in 1982) and very much a part of the development control process.

Two major courses of action are therefore open to developers in their interaction with central authority. Appeals may be lodged and/or representations may be made to influence the broader policy decisions.

There was a reluctance on the part of many housebuilders in our sample to become involved in attempts to influence policy. For most of the small operators and even for some of the large companies, the overriding concern was the day-to-day handling of development control issues. To influence policy, it is necessary to gain access to the higher echelons of the planning authorities and, indeed, to have considerable knowledge of the planning system. Ill-considered attempts to alter policy could backfire.

Should a developer wish to influence policy through the medium of central government, approaches may be made to the officers at the DOE, the relevant ministers at the DOE, or MPs generally. No one in the Central Berkshire sample claimed to have contacted the officers at the DOE to discuss policy. However, as has been noted, the HBF maintains regular contacts with DOE officers and are certainly influential in presenting the housebuilders' viewpoint. The appointment of Tom Baron, the secretary of the VHBSG, as Heseltine's housing advisor speaks for itself. It is clear that he had a very substantial role in the drafting (and by its tone, probably in the writing) of Circular 22/80. Generally, however, contact with ministers is very difficult to monitor. It does occur, but at very high levels within companies and only a few of those interviewed were able to indicate in the vaguest of terms their nature. A few admitted to writing directly to the Secretary of State (about the speed of processing of planning applications, for instance) but whether such documents ever receive the full attention of the minister must be open to doubt.

More direct and effective perhaps is the business lunch at which the minister may hear at first hand the problems of the industry. Undertaken irregularly by only the largest companies, these meetings were used to communicate the problems of the industry. Discussions were of general issues and appeared not to refer to specific cases. The impression conveyed was that Michael Heseltine, then the Secretary of State, was amenable to approaches by housebuilders, who appreciated this accessibility and applauded his policy decisions but were disappointed in the policy implementation. The importance of such meetings must vary according to the personalities in power but there can be little doubt that developers felt they were able to 'bend the ear' of Heseltine.

The role of the local MP appears to be fairly insignificant in the lobbying process. None of those contacted in Central Berkshire claimed to have contacted the appropriate local MP. This may be a local phenomenon resulting from the stance of local MPs against the structure plan modification but is probably widespread. In recent years the HBF has exhorted its members to write to their MPs (especially about land availability issues) but this seems to have elicited little response. The general feeling is that concentration of power within the DOE is high and that approaches to MPs are an inefficient way of getting results.

While there may be minimal direct contact between housebuilders and DOE officers there is very considerable indirect contact necessitated by the appeal system. The appeal mechanism is very widely used: between 1974 and 1981 in the three districts of Central Berkshire, 16.7 per cent of all refused new dwelling applications were taken to appeal. Of those interviewed, 82 per cent had lodged at least one appeal on residential matters (another 5 per cent had lodged only non-residential appeals), although not all had lodged an appeal in Central Berkshire.

The appeal mechanism is therefore widely regarded as a standard and conventional element of the planning system. While the appeal clearly has aspects of confrontation, the frequency of its use means that it need not sour relations between applicant and local authority. The appeal is very much a part of the development game but is used in different circumstances by different applicants.

The use of the appeal clearly separates the very small housebuilder from the rest: less than half (44 per cent) of the very small firms had ever lodged an appeal, whereas the figure for the larger operators was 85 per cent. By virtue of the size of site upon which they operate and their limited resources it is seldom that the very small housebuilder finds it necessary or desirable to appeal; appeals can be expensive both in terms of direct costs to legal advisors and indirect costs in the form of payments of interest rates on the land whilst a decision is awaited.

A number of appeal strategies may be identified, but first it is useful to distinguish between the two types of appeal: the written representation and the local or public inquiry.

The *written representation* is the most common mode of appeal. In 1981, nearly 80 per cent of all appeals were conducted by written representation involving submission of written material. This mode is encouraged by the DOE and by the local authorities because it is quicker and requires fewer resources. For the developer it can be very cheap because it need not involve legal representation, but likelihood of success is low. In 1981, the percentage of all written representation appeals was 30.9 per cent while the equivalent figure for the local inquiry was 39.8 per cent. The lower success rate of written representations may reflect the fact that the less hopeful appeals are dealt with by this method, or that this method is in fact a poor one and carries little weight with the DOE. But to appeal by *local inquiry* involves much higher costs – legal fees for a one-day hearing were quoted as varying between £3,000 and £6,000. The appellant must therefore weigh the extra costs of the inquiry against its increased likelihood of success.

Predictably, it was the larger housebuilders who made greater use of the inquiry (see Table 3.16); their financial resources and the significance of their sites encourage its use.

TABLE 3.16: Number of housebuilders who have appealed by Local Inquiry

	Local inquiry used at least once	*Appeal by written representation only*	*N/A*
Very small	–	4	–
Small	1	7	–
Medium	4	2	–
Large	9	1	2
Total	14	14	2

While one developer claimed to work through the appeal system in progression (first the written representation, then if unsuccessful the local inquiry, and finally, if justified the High Court), others were more pragmatic and identified certain issues as being more effectively handled by the inquiry. General issues (usually involving outline applications) were often thought to be best dealt with at inquiry, while specific issues could be forcefully stated through the written representation.

On one issue all appellants were clear: cross-examination at the inquiry was the one sure way to capitalize upon those instances in which a planning officer's recommendation had been rejected by the committee.

In general terms, therefore, policy issues have been broached with central government by the housebuilders' organizations and by a few influential executives of the major companies. Structure plan and local plan inquiries have been attended and influenced on the housebuilder's side principally by the HBF. The individual housebuilder has benefited from this by the alterations to local and national policies generally and by their influence upon appeal decisions in particular.

Contacts with the county planning authorities

Because of their peculiar role in the existing three-tier planning system, housebuilder contacts with the county planning authorities are limited. As indicated, most residential planning applications require some liaison with the county highway authorities but generally these links do not appear to be of great significance in the overall context. Such links were not extensively investigated.

For a few developers, however, contacts with the county authorities can be important because of their role in policy formulation. In the course of the search for the land for the extra 8,000 houses, the County Planning Department received many 'offers of land'. Landowners and developers, often through their agents, sent information usually in the form of planning documents in an attempt to demonstrate the suitability of their land for residential development. According to the Agenda of the Berkshire County Council Environment Committee (14 December 1982), most of the land proposed for development fell within the areas already designated by the county as suitable sites but much of the information provided was site-specific and highly detailed. This information was made available to the district councils to aid their eventual designation of specific sites and emphasizes the extent to which there is overlap between the work-sphere and style of planner and developer. This is not coincidental: to achieve their objectives many developers have recognized that it may be beneficial to follow the approach of the planning system and from this the not infrequent notion was that the developers were 'doing the job of the planners'.

The county authorities are also used in the search for sites but this type of approach seems only to be used by the larger, non-local developers as a preliminary stage in the land search.

On a more general level, there are contacts between the county and the developers about land availability. The studies requested by circulars have been organized by the counties.

In summary, then, any contact between developers and county planning authorities tends to be on a fairly general policy basis (with the exception of the site-specific land availability studies) and tends to involve developers in the large, medium and the upper end of the small-size categories.

Link with the district planning officers

To begin a development, every housebuilder requires planning permission for the site. Planning applications must be submitted to the appropriate district authority and while in theory there need be no more contact than the passage of application and decision (as would be the case in purely regulatory planning), in practice there is significant contact between applicant and the planning officers. There may be contact with the various technical officers of the authority, but in this section it is the planning officers who are the main concern.

At the head of each district planning authority is the Chief Planning Officer (CPO) and his/her deputy. For large developments, contact with the CPO may be necessary but few developers will request or be asked to see the CPO. (As a scare tactic, the confident or aggrieved developer may demand to see the CPO but in general such demands will be resisted by the department.)

Generally, the planning departments of district authorities are made up of policy and development control sections. Policy is concerned with the more general, long-term aspects of planning and only the larger and more sophisticated developers have any significant contact with the policy officers. But for all housebuilders, the development control offers are of immediate and vital significance. It is these officers who will process the application which will permit or restrict development.

Again, depending upon the nature of the inquiry, contact will be with the principal officer or the area planning assistants (the responsibilities of the planning assistants are usually demarcated by area). The contacts may have several purposes. They may be of an informative, advisory, decision making or monitoring nature but often a contact established for any one of these purposes will encompass others. The contacts will be by letter, face-to-face meeting or telephone (in approximately ascending order of frequency, but with the personal meeting being the most significant). Thus, a relationship is built up between developer and planner as each observes the actions and reactions of the other over time.

Many developers have a clear notion of which authorities they can operate with best, but others claim that it is not the authorities but individual personalities which determine good working relationships. Doubtless, the most accurate picture combines the two: that certain authorities have particular styles which are reproduced in the type of officer which they appoint. However, there was a tendency for the larger housebuilders to speak in terms of the local authority style and for the smaller housebuilders to speak of individual personalities. Such a differentiation may reflect the more frequent and diverse association required by the larger developers; the relatively small and narrow concerns of the smaller builder may mean that the small operators have contact with only the area development control planning assistants, with whom relationships may flourish or stagnate (there being little scope for recourse to other officers).

As a result different developers have different perceptions of the various authorities. Only one authority (not in Central Berkshire) consistently elicited a negative response. A common criterion upon which authorities were judged was their willingness to negotiate (or 'to discuss matters', in the terminology of the interviewees). Generally, the more sophisticated the housebuilder the more he preferred negotiation; the less sophisticated were often irritated by the vagueness of the negotiation. As one small but highly sophisticated operator said, 'I like that local authority because it is prepared to negotiate. But it is not soft and I don't always get what I want – but at least they are prepared to talk.'

This distinction between those who like to negotiate and those who prefer the regulatory, clear-cut aspect of planning is fundamental to an understanding of the criticisms of local authority planning. The characterization of planners as negative and obstructionist came most frequently from those who failed to understand that planning could be a system of negotiation. They complained that too often there was a willingness to object to aspects of a development but no subsequent suggestions as to what might be acceptable. In contrast, the experienced negotiator comes armed with a combination of proposals and tries to elicit just how much he can extract from a site (in the late 1970s, this approach reached an extreme whereby a number of proposals were formally submitted in the form of planning applications – currently, the trend is to discover the possibilities more subtly through pre-application discussion).

The pre-application discussion is perhaps the most important of the contacts between developer and planner. Every developer in the sample regularly initiated this type of contact and only on a few occasions were they deemed unnecessary because the proposed development was so uncontroversial. In Central Berkshire, all authorities were willing to discuss applications before their submission but except on the large sites it was the developer who had to request the meeting. These meetings appear to be of crucial

Block 7 How to get past the planners

A member of a planning consultant's firm operating in Central Berkshire offers an eight-point plan on how to succeed with a planning application.

Information and knowledge: keep abreast of complex and constantly changing planning law and plans, and be aware of whether or not policy objectives are being met. If not, does it represent a change of attitude on the part of the local authority?

Timing: it is critical to know when to submit an application, which follows from information and knowledge. In Central Berkshire for example, despite the requirement to find land for an additional 8,000 houses, applications are being refused as being premature pending the outcome of technical studies to identify suitable locations.

Plan making: get involved in planning policy formulation at EIPs and Local Planning Inquiries. Land allocations can be secured by such involvement.

Quality of the application: should be comprehensive and provide the rationale for the development, the servicing arrangements and the means of implementing the proposal. This involves the preparation of development briefs, management plans and the drawing up of legal agreements.

Development briefs: should be submitted to the local authority prior to the formal application, as this makes your intentions clear and can form the basis of discussions.

Political considerations: know the political composition and politics of a local authority. Too close an affiliation, however, could mean that an application is subject to very close scrutiny. The influence of parish councils on local members should not be forgotten.

Appeals: should be avoided if possible as there is only a 30 per cent success rate, which falls to 10 per cent for major schemes. Written representations are cheaper but are generally 'not worth the paper they're written on' as evidence cannot be tested face-to-face. Inquiries provide this opportunity. Therefore a good QC and team (of planning consultants) are needed. If the appeal should fail (despite the good team of planning consultants), nearly all decisions can be taken to the High Court on a point of law.

Planning gain: has to be relevant; don't offer open space if the local authority is not looking for open space.

significance because although the planning officers can give few firm commitments, the applicant can discover what the planning authority is likely to accept on a particular site. Discussions can be very detailed (the more so on controversial sites) but usually the applicant tries to test the limits of what might be acceptable.

For the major sites with infrastructure agreements to be settled, the pre-application discussions take on a much greater authority. They can extend over a period of years and involve negotiations over legal commitments. Planning gain agreements are frequently drawn up at this advanced stage.

Overall, the rapport between applicant and planner seems quite strong and has doubtless been strengthened by increasing contact. But developers often feel frustrated by the fact that they are not 'negotiating with the people who take decisions'. The power of the committees which ultimately take the decisions is resented and there can be little doubt that nearly all those interviewed would like to see the powers of the officer increased.

Contact with district councillors

Of all the groups or individuals with whom the housebuilder comes into contact, it is the members of the district council planning committees for whom the greatest venom is stored. The more lucid branded the committee meetings as 'Vaudeville' or the 'Muppet Show'. Disrespect was widespread and one of the more sympathetic comments referred to the councillors 'sitting on the fence with their ears on the ground'.

The most cited complaint of the planning system was the situation in which a planning officer's recommendation for approval was overturned in committee by the elected members. If this happened on the grounds of design or detail, the felony, in the eyes of the applicant, was compounded. There was barely an interviewee who failed to voluntarily cite such an instance. A few claimed that it happened to their own applications quite often but others admitted that it had never affected them personally. In the language of prejudice, the willingness of members to overturn officer recommendations was a well-recognized trait. The strength of feeling about this issue prompted an analysis of the frequency of the overturning of recommendations (Fleming and Short, 1984). Admittedly, the work was undertaken in what developers recognize as a 'clean' planning authority (i.e. one in which decisions are thought to revolve around planning rather than political considerations), but the level of recommendation rejection was less than 4 per cent. Clearly, this is a topic on which developers like to swop anecdotes and claim persecution. However, further questioning often reveals a more sensitive attitude: a sensitivity to the desire for some democratic control over development but a frustration by its practice; and a sensitivity to the way in which the role of the committee can curb corruption (the wider the spread of power, the more expensive and hence the less efficient the technique of corruption). Also evident was a dilemma between the housebuilder's business activities and personal philosophy.

The frustration of the developer arises from the amount of time and resources necessary to come to an agreement with the planning officers which can be thwarted by a few minutes' discussion in committee. The relative powers of the officers and members in various local authorities is well understood by experienced developers. Reproaches are often made against officers who have failed to 'fight' for the acceptance of their recommendation and against committees which try to impose their 'political will in planning affairs'. The general impression given by developers was that the local authorities of Central Berkshire were by no means the worst to deal with in terms of their office/member relations. None the less, distinctions were made which characterized Bracknell DC as a 'clean' authority (i.e. where planning considerations outweighed political wishes) and Wokingham DC as a 'dirty' authority (i.e. where the committee's political will was strong). The frustration of the developer therefore stems from the double set of rules which must be played (planning and political) and the inaccessibility of the members and therefore the political rules. But to a large extent this inaccessibility and isolation is self-imposed.

There is a distinct reluctance amongst developers to approach committee members. To a certain extent, this is a result of the lack of a sophisticated approach of some developers (usually the smaller ones) towards planning problems. But it is also a reflection of the image of local government. At the time of its reorganization in 1974, local government was frequently portrayed as inefficient and corrupt. Even though the image has improved, the legacy has a marked effect upon current operations. Whereas councillors may be prepared to meet with developers, the latter are shy to establish contact for fear of backlash reactions or accusations of corruption. Agents, especially, were reluctant to contact councillors (deeming that type of approach 'unprofessional' and 'inappropriate') and where they thought such an approach would be helpful they informed their client accordingly.

Should a developer decide to contact a councillor, there are a number of possible approaches. Most common is the letter to the chairman or to the entire planning committee. Members seldom deem it necessary to respond to a circular letter, and the individual letter is more likely to elicit a reaction. Generally, the letter is the commonest and safest but least effective mode of contact. The more personal contacts by way of the telephone or meeting are more time-consuming, potentially riskier but can be most successful. The personal meeting allows an exchange of views but leaves the councillor more open to allegations of misdealing. In fact, very few developers even attempt to set up meetings with councillors and one who did was somewhat upset when the whole committee turned up. Speaking with the members *en masse* was not considered to be productive. In contrast, a group of large developers from the extensive Lower Earley housing development banded together to arrange a meeting with the planning committee of Wokingham District Council in an attempt to inform councillors of the problems faced by developers.

The issues which developers discuss with councillors tend to be application-specific. It would appear that general policy is seldom discussed (such discussions are aimed solely at the officers) and that the developers' meeting with the Wokingham councillors was atypical. It arose because of the difficulties developers had in obtaining detailed planning permissions, often with increased densities, for that particular site.

Generally, the approaches to councillors are undertaken by the small and medium-sized local developers and take two quite distinct forms. One approach sees the committee member as a political lobbying target to bypass officers' planning-based recommendation. Rather than persist in negotiations with planning officers to secure a positive recommendation, a few (often naive or headstrong) applicants will lose patience and attempt to obtain a favourable decision by making a direct approach to the decision makers. In fact, approaches are made to only one or two members – some developers perceive 'dependency' links amongst certain members and realize that the support of one will entail the support of others. However, this approach tends not to be very successful. Only rarely is a new dwelling application approved against officer recommendations. Residential policies in Central Berkshire are quite clearly stated and officers fight to maintain the status of these policies.

The second approach is much more sophisticated and sees member contact not as a bypass mechanism but as a traffic light. As already noted, members will on occasions reject officers' recommendations for approval. Rather than go to appeal, a developer may choose to try to persuade the committee of the acceptability of the proposal. One example should suffice: a small housebuilder had secured the officers' approval of a detailed application for a site but, at committee, the elected ward member spoke out against the scheme and instigated a member-inspired refusal. The developer did not wish to go to appeal nor to substantially alter the application, so, with the officers' blessing, he made a direct approach to the ward member. After some persuasion and very minor modifications, the member agreed to support the application, and in fact spoke in its support at the next planning committee meeting. The application was passed. This approach was highly sophisticated and, as the applicant said,

> 'By contacting the councillors I am doing part of the planning officers' job. But I would not do so without their consent. My approach is to secure the officer's support and only then will I consider other methods to secure the passage of my application. That way I gain credibility in the eyes of the officers and the members. It takes time but it makes my job easier.'

None the less, it should be stressed that developer-councillor contact is as yet uncommon. But, as developers become more sophisticated and planning-conscious, it is likely that such contact will increase and that developers will seek to work with the planning officers rather than to bypass them. From interviews with district councillors, it is apparent that few members find developer contact unacceptable, although many treat them with scepticism. On the developer side, future contacts can be expected to be more forthcoming from the more sophisticated larger housebuilders.

Contact with parish councils

The influence of parish and town councils in planning matters is often overlooked. They are one of the few statutory consultees of development control and policy matters. They receive notification of all planning applications in their area and are invited to submit their observation to the district council planning officers. They also make their views known to their district ward representatives with whom their views carry considerable weight.

Although their powers do not even reach the status of a 'recommendation', their interest in planning affairs and their influence upon others might be expected to make them the objects of lobbying by developers. To give an indication of the interest of parish councils in planning matters, the circumstances of twenty planning applications at two meetings of the Bracknell District Council Development Committee may be noted. Eleven received parish council comments: they suggested refusal for five, approval for four, asked for extra conditions on one, and offered 'no observations' on one. Of the five to which they objected, two were refused (one against officers' recommendation), and the remaining three had additional conditions imposed.

In opposition, or in support of development, therefore, the views of the parish councils carry significant weight. Unfortunately the extent of this influence did not become fully apparent until the interviewing was well under way and no direct question concerning the parishes was included in the housebuilder's questionnaire. None the less some information was elicited and later supplemented by the questionnaire of parish council members.

Lobbying of parish councillors by developers is not common, and is generally accompanied by an approach to the appropriate district councillor. While there is a reluctance to contact the latter because of the fears of the 'inappropriateness' of such action, there is a reluctance to contact parish councillors because they are presumed to carry little or no influence. However, when contact does occur, it appears to take two different forms.

First, there are the personal approaches made by the smaller housebuilders. These housebuilders generally lived locally and made themselves known to the parish councillors, and were building up high levels of local credibility. They made their approaches just prior to the submission of their applications for small, occasionally contentious, infill sites, by securing the confidence and support of the key members of the council, aided the safe passage of their applications. Lobbying was most often undertaken by up-market developers (though perhaps trying to secure a fairly high density) who would impress upon the councillors the suitability of their style of development.

Second, there are the more formal approaches, usually made by the larger housebuilders or their representatives to meetings of the parish council or its planning committee. Such contacts were undertaken by the most sophisticated operators and although the contacts tended to be persuasive rather than negotiationary in nature, there were occasions when the contact was genuinely two-way. One parish council ultimately secured a village hall, through supporting a particular residential development.

Contacts with community groups, residents' associations and neighbours

Direct contact with community groups or individuals was relatively rare but difficult to establish with certainty. Often only one member of staff in each company would undertake such contacts and so, especially in the larger firms, interviewees were often uncertain as to the frequency and nature of the links.

The public groups with which developers might have links may be identified as of four types. First, there are the 'management committees' initiated and often given some capital funding by the developer. Essentially, these committees are associations of the purchasers of the new homes and are encouraged by the developers to take over the management of flats sold on leasehold or of the public open space left on the new estate. These management committees are, to a certain extent, extensions of the development industry. In one sense, the developer is subcontracting out work: the developer's input is initial organizational aid and perhaps a once-for-all payment; the work, however, is subcontracted out for all time and the developer need retain no responsibility for it.

Second, there are the incorporated groups. These are generally public groups which have been aided by the planning system to have some voice in the development of the large estates. Usually comprising consultative committees, the developers sometimes have direct contact with them. Too late to stop development (though sometimes originating in an attempt to do just that), their role is to try to have some say in the details and progress of the estates. Secure in the knowledge that the principle of the development is safe, developers are often happy to participate and sometimes to modify their plans or methods of working as appropriate. No developer claimed that major change had been instigated by such contacts and many saw their own role as being educative, informative and thereby persuasive.

Third, there are the community groups which establish themselves to fight specific proposals for development. Often they try to fight the principle of development in a particular area but by the time many developers learn of this, contact would be futile because there would be insufficient time to act. Some developers with a sound local knowledge did initiate such contacts but this was very rare. Two interviewees considered that perhaps they should make efforts to gauge local opinion, but for the most part the few contacts which were reported were initiated by the groups. Developers' responses varied but generally took one of two forms: the contacts were a nuisance, achieved little, were largely irrelevant and avoided where convenient, or the contacts were of limited value, their value being that the developer could be seen to be 'responsible' and that this could be used as evidence in any subsequent appeal to support the case for development. Clearly, a few developers realized the importance of allowing as many people to play the development game as possible even though the developer considered that the bat (the housebuilding materials) and the ball (the land) were his own. A unanimous complaint of the developers was that these new players were ill-informed and understood neither the rules nor the techniques of the game.

Fourth, there are contacts with the neighbours of a proposed or arising development. These evoke quite a different response from developers, though it is only the very small and small housebuilders who get involved at this level. This is probably because a neighbour response is only encountered when dealing with very small sites. While small developers may try to avoid contacts with groups, they are confident in dealing with individuals. In fact a few developers even claimed to initiate contacts with neighbours. One went so far as to post plans of the proposed development through the letterboxes of the neighbours requesting that if they were unhappy with the development they should contact him. This appeared to be effective in gaining credibility and trust amongst many communities.

The role of architectural/planning agencies in the planning process

The architectural profession has always been heavily used by the housebuilder. The technical skills of the architect are in demand for the design and drawing of house types for submission with planning applications to the planning authorities. Architectural courses do include elements of planning procedures and planning law but these are not particularly stressed. In their practical work, however, architects have become an integral part of the planning process.

Davis and Healey (1983), in a study concentrating mainly on the larger companies, drew attention to the increasing role of architectural and planning consultancies to private development. It seems feasible that in attending planning meetings with their clients architects have established regular contact with planning departments. From a minor role of acting as technical advisors, architects may gradually have taken on a much more significant role as they gained, from experience, more information about the whole planning process. As anyone dealing with planning authorities will realize, regularity of contact is of enormous importance in establishing a trusting working relationship. The discretionary powers of planning officers is such that the most prized information and help will be retained for the established and trusted contacts.

In our sample just over 40 per cent of all the housebuilders interviewed claimed to conduct meetings with planners by themselves (or their in-house staff). The proportion varied according to the size of housebuilder and it was the large and the very small housebuilders who made greatest use of agencies.

The largest companies are most confident and self-sufficient in dealing with the planners and this is to be expected since half of them employ 'in-house' architects. Only one company who employed an outside agency did not use that agency in their planning negotiations.

The small and medium-sized housebuilders seldom undertook discussions with planners alone. From the discussion of the land search, it might be expected that these local builders would have sufficient local knowledge and frequency of contact with planners to enable them to undertake their own negotiations. They may have the ability to do so but they recognize the benefits of utilizing their architectural consultants. It is again significant to note that only one who used an outside architect did not utilize their services in the planning sphere.

The agent is the interface between private development interests and the planning system. One architect described his role in terms of doing his 'best for the applicant and society at large'. No contradiction was apparent to the architect. Clearly, they have considerable, subtle power and, in Central Berkshire at least, their values and mode of operation are having an enormous impact upon development.

Their impact also seems to be growing and in Central Berkshire one local agency is having considerable success, having recently established a specialist planning consultancy section which alone operated for eight of the interviewed developers (its list of clients numbered thirty-three). In addition, this practice also served for other developers in a more specific architectural role. Its services are geared towards the large housebuilders who have large financial resources but are, perhaps, lacking in local knowledge. The success of this consultancy is so great that even a few small and medium-sized firms have engaged its services. It is difficult to predict how long a single agency can be successful operating for several housebuilders in one area, particularly when one of their services is a land search. To date, that problem seems to have been adequately negotiated by the consultancy.

A summary of the approaches of housebuilders to the planning system

By the nature of the development process, every housebuilder must interact with the planning system. Some housebuilders deliberately keep such interaction to a minimum whereas others invest considerable resources in attempting to obtain a favourable outcome. We have already noted a split in housebuilders which indicated that some concentrate on production to obtain favourable returns upon capital (the 'constructors') and that others stress the importance of bringing forward undesignated land for residential development (the 'landfinders'). The latter group need to have considerable interaction with the planning authorities whereas the former need not.

In this section, some parameters and a typology are forwarded to try to give an impression of the quality and quantity of housebuilding companies' interaction with the planning system.

One readily available parameter which can be used as a very crude indicator of a company's success in dealing with the planning authority is the proportion of their applications which are approved (see Figure 3.4). The majority of the sampled companies achieve a success rate of more than 70 per cent. Surprisingly, a high proportion (principally the very small and the large developers) enjoyed a 100 per cent success rate.

Cutting across the successful rates is another parameter (though exceedingly difficult to measure) which might indicate the degree to which an applicant probes and explores the limits of planning. This parameter has been assessed on a purely intuitive basis as a result of impressions gleaned from the interviews. By exploring the limits of local planning, the developer attempts to obtain the most favourable outcomes from the system without incurring a high rate of refusal. Combining these two parameters a matrix can therefore be envisaged which identifies success rates and the probing of the limits of planning (Figure 3.5). From this, a fourfold typology may be conceived.

The *cautious* housebuilder has a high application success rate

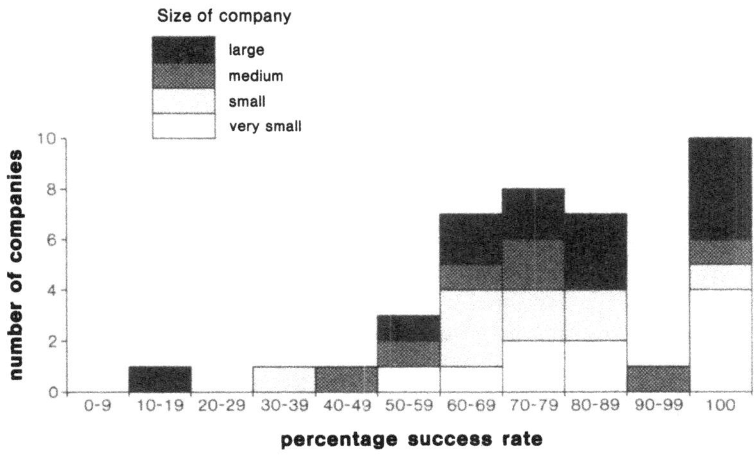

FIGURE 3.4 *Planning application success rate of housebuilders in Central Berkshire 1974–81*

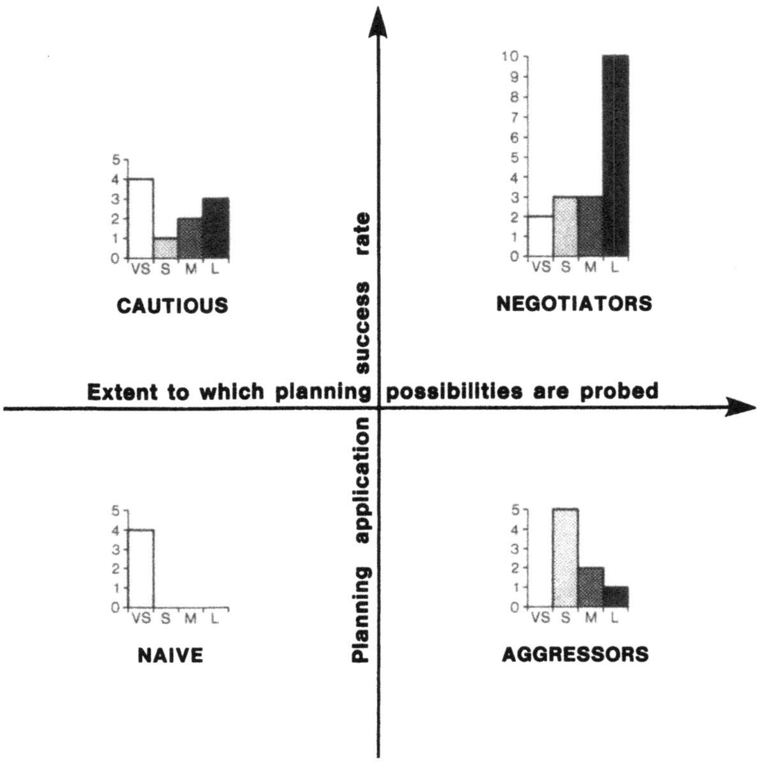

FIGURE 3.5 *A typology of housebuilders' approaches to the planning system (numbers refer to very small, small, medium and large firms)*

but involvement and interest in planning matters is low. Few surprising decisions are therefore encountered. The conservative style of these housebuilders restricts them to safe sites which often have outline planning permission even before they are purchased. Contentious applications are seldom submitted because the housebuilder usually wants quick, approving decisions with a minimum of planning interaction. Many companies which concentrate upon production are included in this category.

The *naive* housebuilder is rare and, almost by definition, restricted to the very small, often one-man operated firms. They have little interaction with the planning system but because they are somewhat ignorant of planning procedures, they suffer a relatively high level of application refusal. Through naivety, they submit hopeful or careless proposals which are unacceptable to the planning authorities. Even when faced with refusal their lack of knowledge of the system precludes a negotiated compromise (unless local planning authority-initiated) and a lack of funds or confidence often discourages the submission of an appeal. Besides, the applications are generally refused either because the proposal is totally inappropriate in which case the proposal dies, or is unacceptable on grounds of detail which may be relatively easily amended on a subsequent submission.

Only the small firms can stumble from one site to another in such a naive fashion and yet survive. Larger firms have to support larger organizations and either quickly become cognizant of their failings and take corrective action or they topple under their own weight.

Both the naive and the cautious developers have minimal contacts with the district planning officers, and probably no contacts with the perceived periphery of planning powers (the elected representatives and the DOE in the form of the planning inspectorate). Only rarely will the cautious developers employ planning agencies and even then the agencies are either of a cautious type or are given a very restricted brief.

The *negotiators* stand in stark contrast to the naive and cautious developers. They have a relatively high application success rate and often their interaction with the planning system is such that they have virtually become incorporated within it. In this context, planning is not a regulatory instrument but an arena for negotiation ranging from pre-application discussion to (more rarely) formal appeals. Negotiation is sometimes undertaken by planning agencies. It is no coincidence that many of the quality negotiators have emerged from the ranks of the local authority planners. They understand the game, and accept its rules, but through their position as representatives of private capital they explore, bend and test the limits of planning. At its most sophisticated, there is considerable rapport between negotiator and planning authority and a very high degree of respect.

The sites with which the negotiators are concerned are variable. They are by no means necessarily safe development sites. Negotiation may span years and be aimed at making the local authority realize that a site is appropriate for development. On the other hand, many of the smaller negotiators may concentrate on securing relatively high densities. The trade-offs are many and various.

The negotiators, then, encompass the full range of housebuilder size groups and planning gain agreements are very much the symbol of this group of developers.

Finally, the *aggressors* are those with a high degree of planning knowledge but none the less a relatively low success rate. But that is not to suggest that they are unsuccessful negotiators. They may be impatient negotiators or a little too ambitious in trying to obtain permission for outrageous proposals, or they may be more eager to defer to a higher authority in the DOE rather than the local planning authority. Their interest in negotiation is therefore variable. The line between the aggressors and the naive developers may be fine but the resource and ability to take a decision to appeal means that the arbitrariness of appeal decisions may be fully exploited.

Aggressors usually operate most effectively when outline planning permission is at stake. Therefore, many of their number tend to be those who profit from the land rather than production. A substantial proportion of the aggressors were from the small size category (although some of these bordered closely with the naive developers). Their preponderance in this category may reflect that it is they who are feeling the pressures of the large companies in Central Berkshire – their stance may be something of a desperation measure in an attempt to obtain suitable small, but not single-plot sites. Under the Heseltine regime, with its shift in central policy directives to a pro-developer stance, this group was given considerable confidence.

It is the negotiators and the aggressors then who invest most resources in tackling the planning system. It is they who are actively involved at the local level and who have given impetus to the assault upon the system undertaken at the top through individuals and the housebuilder organizations.

Emerging trends in housebuilding

The 1970s witnessed major changes in the housebuilding industry. Many of the trends have already been alluded to earlier but four major ones will be reiterated in this final section in an attempt to pull together some general themes.

(1) Increased politicization and professionalization have been common characteristics of many industries and organizations over the past decade. They have taken particular forms in housebuilding. While increased owner-occupation has been a central plank of every post-war British government's policies, the public profile of the new home has, until quite recently, been remarkably low. Until the mid 1970s the politicization of the housing issue was implicit and taken for granted by the industry. The general political climate of the early 1970s aided the rise of the land availability debate but it was not until housebuilders perceived the failure of structure planning in the mid-1970s that any real

pressure was mounted by the industry itself. Led by Tom Baron, the volume housebuilders initiated a major lobbying campaign. Monitoring the gains which were being achieved for major builders and not wishing to be left behind, the HBF revitalized itself and launched a major assault upon the prevailing planning system. Feeling left out the FMB and other smaller organizations also joined in. And so a high-profile politicization of the housebuilding industry has been set in motion. There can be little doubt that the industry has succeeded in bending the ear of successive DOE secretaries of state and in the latter half of 1983 the Green Belt debate had been revitalized (see Short *et al.*, 1984) whilst it was clear that builders' demands for more land were being sympathetically handled. None the less, the industry has been acutely aware of the anti-developer lobby and much of the industry's venom has been directed at that lobby (see, for example, the Directors' Column in *The Housebuilder* throughout 1983). The industry considers that its own increased politicization has been in response to the rise of the anti-development lobby.

The increased politicization shows no signs of waning; depending upon government action, the process may in years to come appear to have peaked in 1983 but at this time it is difficult to judge.

Much of the politicization has focused upon the issue of land and, indeed, housebuilders' attitudes towards land have changed markedly. From the depths of a recession, the prevailing attitude has been not to hoard land through purchase but to make it available for future use through low-cost agreements. The trend has been away from the maintenance of land banks and the acquisition of profits from windfall land planning gain to the safety of short-term serviceable land banks and long-term agreed supply. The current movers and shakers of the industry are those who concentrate upon production.

(2) Whilst Tom Baron has been the figurehead (and more) of the attack upon land availability and planning issues, it has been Lawrie Barratt who has undertaken that role in the marketing of dwellings. He has replaced the old 'build and sell' adage by the new 'sell and build'. The phenomenal rise of Barratt housebuilders from a relatively small operation in the early 1970s to the undisputed biggest builder of the early 1980s has highlighted the relative backwardness of the industry as a whole. That one company should make such a phenomenal rise has served to indicate the complacency and conservatism of the industry as a whole.

Although Barratts have undoubtedly led the marketing revolution, the company has merely exploited rather than innovated the trends in production. Production trends are typified if not dominated by the timber frame approach. Cutting direct employment to a minimum, companies can trim their labour responsibilities and reduce the production time and costs of housebuilding.

(3) There has been the rise of the privatized planner; architectural/planning agencies (in Central Berkshire at least) have taken on an ever-increasing role in the housebuilding process. From the design of dwellings, their influence has extended to the whole planning complex and is extending even into the field of land finding. The 'privatized planners' bring sophisticated skills to housebuilders, but as their power increases, so does their ability to influence the form and location of development, and the approach of their clients. In their hands negotiation has become an art form.

In contrast, there would appear to be a decline in some of the traditional agents of housebuilding. There are signs that the influence of estate agents in land finding and the sale of new dwellings is on the wane. Increasingly, many of the major builders have established in-house (or pseudo-contracted) agencies to deal with marketing matters. However, some of the larger estate agents have reacted and made their own operations more efficient. It is quite possible that the larger estate agents will market more of the new houses, while the rest will devote their attentions to the second-hand market.

(4) Finally, the trends within the industry unquestionably point to the increasing dominance of the major public limited companies. Whilst the small local builder may survive on small sites, and whilst some medium firms may retain a share of the market by producing high quality (though not necessarily high-priced) dwellings, it is the very large companies which are taking a larger share of the market. So far they are succeeding – at the expense of the small to medium-sized companies.

Throughout this chapter we have constantly referred to the relationship between the housebuilders and the local planning system. In the next chapter we will focus attention on this system.

4 The local planning authorities

'the strange mixture of reality and illusion, democracy and privilege, humbug and decency, the subtle network of compromise, by which the nation keeps itself in its familiar shape'

(George Orwell)

Because the local planning authorities have a statutory duty to prepare land use plans and implement these plans through the development control system, they can become the focus of conflicting pressure between developers seeking planning permission and the various groups attempting to resist or deflect development. In this chapter we will switch our attention to these planning authorities, focusing on the constant negotiation and decision making of development control rather than the more irregular process of structure plan preparation. Those interested in obtaining an insider's view of this latter process in Central Berkshire are referred to Stoddart (1983).

Local planning authorities consist of two sets of agents: elected councillors and salaried officials. These two groups, often referred to as the members and the officers, work together but have different organizational structures, perspectives, priorities and roles. While the officers have to serve their elected members, they also have reference to an ideology of professional planning practice (Healey and Underwood, 1977; Underwood, 1980; Knox and Cullen, 1981). The members, in contrast, especially planning councillors, have to meet directly the competing pressures of central government directives, local pressures, housebuilders' pressures and residents' group demands. The decision making ultimately lies with the members. The competing pressures of accumulation and legitimation noted in Chapter 1 are crystallized in and through the action and role of members. In this chapter, then, our concern with the local planning authorities will concentrate on the members. Before considering the development control system in operation let us consider the characterstics of the planning members in Central Berkshire.

The councillors of Central Berkshire

Central Berkshire's population of 375,000 was in April 1983 represented by 162 district councillors and forty-five county councillors. Elected at various stages over the previous four years, a large majority of the councillors were members of the Conservative party. The Conservatives held healthy overall majorities in three of the districts but in Reading Borough Council they had to rely on Liberal support and on Berkshire County Council an Independent held the balance of power (see Table 4.1). The 1983 district elections strengthened their position in all four authorities, and gave them overall control in Reading. But who were these representatives? It was members of the planning committees who issued decisions which partly determine the location and style of the built environment. Interviews were therefore undertaken with one-third of the district planning councillors, plus a number of leading county councillors. Understanding of these councillors requires consideration of two major factors; the party political context within Central Berkshire, and the personal characteristics of the elected representatives.

TABLE 4.1: Party affiliation in each of the districts of Central Berkshire and on Berkshire County Council at April 1983

	Conservative	*Liberal*	*SDP*	*Labour*	*Total*
Newbury (East)*	12	6	1	1	19
Wokingham DC	46	7	–	1	54
Bracknell DC	27	–	6	7	40
Reading BC	22	11	–	16	49
Total	107	24	7	24	162
Berks CC	43	16	–	27 1	87

*Newbury (East) consists of only that part of Newbury DC which lies within the Central Berkshire Structure Plan area

The party political context

THE CONSERVATIVES

The Conservative controllers of the shire counties are often typified and caricatured as the landed gentry and the owners of local industry. They are seen as the 'old guard' whose concerns are with the order of society; implicit in their philosophy is the 'natural order of things'. So 'natural' has been their status that they frequently claim to be apolitical and until the fairly recent past large numbers used to stand as Independent councillors. They have a deep sense of place and in planning terms have sought to maintain the identity of their towns and villages by reducing sprawl and the inflow of migrants. As a result, their stance is essentially that of anti-growth.

Approximations to these stereotypes are certainly found amongst Conservative councillors particularly at the county level but the post-war influx of migrants has introduced another important type of Conservative representative.

This second strand is the urban and suburban-based Tory associated with what we may term 'civic boosterism': the promotion of local business activities. They are most prominent in Reading and Bracknell and strive to reduce rates and operate an efficient 'minimalist' local authority. Growth is not an anathema to them, although their support for local expansion is sometimes tempered by political caution. Rather than supporting unlimited growth, they have adopted an intermediate line calling for growth but not at any cost.

Between these two strands of Conservatism there are of course more opportunist Tories. Most of these have moved into the area to take up positions as professionals within the new national and multinational industries of the Western Corridor. They have sought to establish roots within their local communities and to a greater or lesser extent have endorsed the anti-development sentiments of their neighbours.

THE LABOUR PARTY

The Labour party strength in Central Berkshire is almost exclusively confined to the new town of Bracknell and the old urban centre of Reading.

But its presence even in these two areas has hardly been prodigious. During 'good' Labour years in the 1950s and 1960s, Reading was inclined to return Labour MPs. However, subsequent boundary changes have virtually ensured that Reading constituencies will never again be represented by Labour at parliamentary level. In Reading, Labour had overall control of the borough for a short period after local government reorganization, but after almost a decade with a hung council, the Conservatives gained control in 1983.

Labour activists in the area have been of two broad types. There are the 'native' working-class manual workers, often employees of the local service industries (rather than the traditional 'brewing, biscuits and bulbs' industries of Reading) who were originally largely based in the inner urban areas. There are also the professional classes who have been attracted to Reading's expanding economy throughout the post-war period. Often from working-class and trade unionist backgrounds, some have maintained their family traditions of Labour activity. But in their social and locational mobility, some have taken a more radical Labour line which has not always sat happily with the 'native' traditional style, whilst others have adopted a more centrist, SDP-like approach. Throughout the 1970s tensions were evident between the radical and traditional elements.

In contrast to Reading, Bracknell has never had a significant body of traditional Labour activists. As a new town its population came from elsewhere, often London. However, the Bracknell Labour party was subject to severe tensions in the early 1980s between the 'left' which came to dominate the town council and the more centrist district councillors. Conflict reached such a pitch that many Labour district councillors defected to the SDP. The consequent split in the anti-Tory vote in the 1983 local elections enabled the Conservatives to win every available seat on the district council.

THE LIBERAL PARTY

The Newbury-Reading area has in the post-war years been an area of significant Liberal support. Even though no Liberal MP has recently been elected in the area, nor has any local authority had a Liberal majority, the party has been very strong in small areas, particularly to the east and west of Reading.

The strength of the Liberals in particular neighbourhoods is not merely coincidence, nor can it glibly be attributed to a 'neighbourhood effect'. It is very much a reflection of the Liberals' approach to local politics. At their annual conference in 1970, the Eastbourne Resolution was passed which called for the party to accentuate community politics at the local government level. Essentially, 'community politicking' stresses local issues of place and people rather than national issues and Liberal philosophy. Thus, by wooing constituents of particular wards, hard-working community politicians with or without fully developed political philosophies can secure sizeable pockets of Liberal support.

THE SOCIAL DEMOCRATIC PARTY

True to its media image, the SDP is the party of the young professionals. Its very recent emergence makes it difficult to assess but it has made small strides in West Reading (in Newbury DC) where it has had to build a new base in a mainly Liberal area, and in Bracknell town where it has grown through Labour defection. Elsewhere there have been fewer signs of a breakthrough. Of the few councillors sampled none originated from Berkshire or even the South East, and the suspicion is that at councillor level at any rate, the SDP is the expression of a non-southern family background of Labour and trade unionism translated to a growth area in the 1980s.

Councillor characteristics

In 1983 there were fifty-five district planning members in Central Berkshire: nine in Newbury East, eighteen in Wokingham, twelve in Reading and sixteen in Bracknell. We interviewed nineteen councillors in March and April of 1983, including the chairperson of each planning committee (see Appendix 3). More detailed information is given in Witt and Fleming (1984).

Despite insistent protestations from many (but mostly the Conservatives) that party has no place in local politics, let alone planning issues, we consider that in implicit if not explicit terms, party *is* of significance in attempting to understand councillor characteristics in Central Berkshire. Party and district will therefore be important reference points in the following discussion based on the questionnaire survey results.

While the Maud Committee Report (1967) indicated that councillors are more likely to emerge from the older sections of the community, two later studies (the Robinson Report, 1977, and Gordon, 1979) suggested that there might be a trend towards younger representatives. For the Central Berkshire planning councillors, the age-distribution was fairly uniform but there were major inter-party differences (see Table 4.2). The Conservative members were the oldest, a reflection perhaps of the strength of the 'old guard' and/or that younger potential candidates are too committed to career advancement to countenance the workload

TABLE 4.2: Age and political party of surveyed district councillors (+ indicates chairperson)

Political party	*Conservative*	*Liberal*	*SDP*	*Labour*	*Total*
Age					
20s			1		1
30s		1		3	4
40s	2 +3		1		3 +3
50s	2 +1	1			3 +1
60+	3	1			4

of local politics. In contrast, the Labour members were the war children of the 1940s. The Liberals were more evenly spread through the age groups but were supplemented by a younger SDP element.

As befits an area of growth, few were born within Berkshire but most had lived in the area for more than fifteen years. The Conservatives tended to be the longest established, but for all councillors it would appear that even in a growth zone, a period of residence is necessary before either party or electorate are prepared to adopt a candidate. Intriguingly three of the four district chairpersons (all Tories) and the two Conservative county councillors interviewed were born in Berkshire. The suggestion is that at the county level (which traditionally attracts a particular type of councillor) and in the positions of power the 'old guard' dominate, while the rank-and-file consist to a greater extent of the newcomer professionals.

Occupations of the interviewed councillors covered a remarkably narrow band. Three were retired, two were housewives, one was employed part-time and the rest were drawn from the managerial classes. Not a single manual or non-skilled worker appeared in the figures. In addition, the scarcity of councillors who were self-employed or who held directorships was notable. Only two held directorships, thus confounding many preconceptions about the prevalence in local politics of the controllers of local industry and commerce.

In contrast to occupation, educational qualifications were of much broader base. As many as five members had no formal educational qualifications whatsoever, but the largest category (of six) encompassed those with some sort of diploma or professional qualification. The syndrome of the self-made member was strong.

Every study of councillors at some stage emphasizes the under-representation of women – a fact which was also true of Central Berkshire. However, it was an interesting if unexplained observation that although males formed the majority on most committees, there was none the less a relative over-representation of women on the planning committees, an imbalance which was statistically significant (at 95 per cent using χ^2). For some reason(s), perhaps a greater sense of community, women seem to gravitate towards the planning committees. Indeed Wokingham DC's planning committee had a tradition of being regarded as a strong women's committee, and in the electoral year 1983/4 thirteen of the committee were women, including both the chair and vice-chair.

But how indeed do councillors come to sit on particular committees? A councillor usually sits on two main committees and some or all of their subcommittees. Since each council has four to six main (or programme) committees, some form of selection is necessary. To ensure that the council's party political balance is reflected on each committee, allocations are usually organized on a party basis. Several Conservatives reported that in the early days of local government re-organization and before, the Conservative leaders would ask (or tell!) members to sit on particular committees. However, the practice for all parties in the early 1980s was to ask members to rank (on election to the council, and some times repeated annually) the committees on which they wished to serve.

From such an arrangement, it might be expected that particular popular or powerful committees would be oversubscribed. Conveniently, this appears not to be the case. Blondel and Hall (1967) noted that councillors of Maldon and Colchester, when asked to indicate the committees upon which they wished to serve, managed to distribute themselves fairly evenly amongst all the committees.

The rankings by planning councillors indicated that the planning committee had been the first choice of eleven members and the second choice of three. At district level only one councillor appeared to have been 'pushed' on to the planning committee.

The reasons why councillors chose to sit on the planning committees were various. For a few, it was a planning matter which first caused them to become involved in local politics; the high level of office construction accompanied by the low level of housebuilding in Reading was the major factor in the case of one Labour councillor. A general interest in community and place caused many Conservatives and Liberal councillors to opt for the planning committee.

A rather different pattern emerged when councillors were asked to indicate the most 'important' committees; only three ventured that planning occupied this position, with five placing it second or joint second. The meaning of 'importance' in this context is rather complex however. Given a tendency to equate importance with control of the purse strings, planning is seen to offer little scope for *direct* involvement in financial matters. Set against this however it was recognized that planning decisions have important consequences for local communities.

The factor complicating the importance of planning committees is the restriction of district council power and discretion by central government. Planning councillors have few illusions about the power of their committee, certainly after a short period of service. To what extent this causes ambitious power-seeking councillors to look for other committees is unclear. As was noted earlier, central government continues to change the framework for development control decisions through circulars and plan modifications. These 'new rules' can be enforced via the appeal mechanism. Thus at the level of both plan making and individual applications, the local authority often does not issue the final decision. Central Berkshire provides ample evidence of this: the planning committees do not wish to sanction the level of residential growth proposed by central government, yet they are left with little option. None the less, the actual decisions made have important physical and social repercussions. The manner in which councillors attempt to accommodate the pressures of both central government and the local electorate is crucial.

Motivations

Many studies have attempted to classify councillors, based on motivation and/or function, among them Rees and Smith (1964), Hill (1972), Corina (1974) and Gyford (1976). A typology forwarded by Elcock (1982) identifying policy brokers, people's agents, local government careerists and status-seeking councillors appeared relevant to our study, and some elements of this have been adopted. Elcock's typology however utilizes both motivations and roles; we consider it important to make a clear distinction between them. Councillors similarly categorized on the basis of role can have very different motivations, and vice versa. In many cases, of course, strong links exist between motivations and roles, and a cross-tabulation of likely and unlikely combinations will be presented.

The motivation of the *vested interest* is easy to understand but its expression is often difficult to interpret. Councillors may pursue their interests by legitimate or illegal methods. Whereas the pursuit of interests through illegal methods usually implies little other than self-interest, the pursuit of interests through legitimate channels may be undertaken with a sincere belief that the community's interests are also best served in this way. The planning sphere certainly offers opportunities for gain; landowners of specific sites and the 'property professions' in general may stand to obtain material benefit from development. But this is not to imply that everyone with a financial interest will seek to gain through office. On the contrary, some councillors actually shun situations in which it might be considered that they are furthering their own interests. Corruption or accusation of corruption are highly sensitive issues in local government. Councillors in all authorities are expected to publicly announce any financial involvement in, or family connection with a committee item, and to

Block 8 The registers of interests

It is commonly imagined that directors of local industry and the landowning gentry are well represented in local government. Morris and Newton (1970) have noted, however, how the power of big business in Birmingham was superseded by 'professional businessmen' and 'exchange professionals' with interests in the local property market. This rise in the professional businessman has also been noted by Saunders (1979) in Croydon.

The Local Government Act of 1972 required that a register of councillors' interests be maintained by all local authorities. Although a standard form was issued its use was not compulsory. It was used by Wokingham DC, but Reading BC, Bracknell DC and Berkshire CC devised their own forms. In Newbury a register was provided but had not been completed by any of the members. Concern in this district seemed to focus much more upon registering an interest with regard to particular items than in general terms. In the other authorities, a form was sent to each new councillor. Completion is generally voluntary but in Bracknell DC, a standing order made this compulsory.

The registers differed and therefore were not strictly comparable. In the four authorities surveyed (Newbury was excluded), there were entries concerning remunerated directorships and employment of the councillor and his/her spouse. Names of firms were requested but the nature of the business was seldom recorded and the position of the respondent seldom evident. All shareholdings over specified limits were to be disclosed in Bracknell and Reading, whilst in Berkshire it was restricted to companies which might trade with the council. In Bracknell, Reading and Berkshire, entries also concerned land holdings and property ownership – and in Reading an entry specifically requested information about interests in property for which planning permission might be sought. Berkshire also requested information about trades, professions, consultancies and membership of organizations. From the responses, it was possible to construct a picture of the interests of each council (see Table 4.3).

TABLE 4.3: Interests of councillors and their spouses in local authorities of Central Berkshire as at May 1983

		Districts			
	Berkshire County Council	Wokingham	Bracknell	Reading	Newbury
Councillors	87	53	40	45	
Entries	73	24	29	23	N/A
Holding directorships	15	2	5	3	
Holding multiple directorships	8	0	1	2	
Landowners	4	N/A	0	0	
Planning related interests	7	3	2	2	

Block 8 continued

The response rate was highly variable. True, some councillors had recently been elected and might yet notify their interests, but it was clear that the registers were not meticulously maintained and the surprise which met the authors' requests to view the registers indicated the extent to which they were consulted.

While councils were certainly not filled by the directors of local industry, these were none the less well represented. Notably, they were more prevalent on the county council than elsewhere (despite the fact that Conservative representation there was markedly less than in the districts).

The lack of landowners and indeed farmers was a surprise. It appeared that the districts had no major landholding interests – landowners were confined, though in small numbers, to the county. As a shire county, it was expected that the landowning interests would be present in force – they were not. Neighbouring Hampshire CC showed a similar low proportion of landowners although one major landowner held a very prominent position on that council.)

The various development and construction interests who might be concerned with planning issues were present at both tiers of government. However, myths that local builders dominated committees were given no support.

From the various sources of information available, it became clear that councillors were drawn overwhelmingly from managerial groups. Hardly a single manual worker was known to exist on any council; teachers, and computer and electrical industry executives, formed the backbone of most councils.

It is of course dangerous to confuse numerical strength and power. It was reputed, and widely believed by local politicians, that in the recent past the strongly Conservative county council was effectively run by a powerful triumvirate – a landowner, a valuer, and a construction company director.

withdraw from the discussion. Block 8 examines the declared interests of Central Berkshire councillors, as recorded on local authority registers of interest.

The suspicion of vested interests can never, however, be ruled out. A parish councillor whose parish had received a large proportion of the 'Heseltown' allocation expressed in sincere terms his views:

> 'Our meetings with county councillors were helpful up to a point, but our attitude to them has been tempered by our knowledge of vested interest in the Tory party. We are cynical enough to believe that vested interests are at work – there are known estate agents on the county council, the land for the new housing is owned by a former county councillor, and we all know about the activities of . . . who has obtained county council contracts.'

Even if these accusations are untrue, the fact that they are believed to be true has influenced attitudes and actions.

Hard financial gain is one possible benefit to be reaped from office, but much more common are various types of soft gain.

Rees and Smith (1964) noted how, particularly for Labour councillors with unrewarding full-time jobs, a career as a councillor could offer a post equivalent to management status. We shall term such councillors *para-careerists*. In our study (in an area with a high proportion of full-time professionals) there were no clear-cut examples, but we did encounter 'post-careerists': councillors who had offered their services to local government upon retirement. 'To exercise the mind and to contribute to the local community' was seen as a fitting reason to take office. It did, however, involve a measure of self-satisfaction and personal gain. One councillor had sought election a few years prior to retirement in order to achieve continuity once full-time employment ceased.

Seldom admitted but clearly evident in the attitude and tone of some councillors was a feeling of power and status derived from their position. To be a councillor is to raise one's status within the community, and to obtain a senior post, however ineffectively executed, can be a considerable ego-boost. *Status-ticians* can be considered to be indulging their egos at the expense of the public, but it is doubtful if many pure status-ticians exist; their role playing soon becomes apparent to their peers who will act to remove them. But as a facet of most councillors, a degree of self-importance is likely to be significant.

Also indulgent of personal considerations are the *social councillors*. They can be found in all parties and their common trait is to treat the council (occasionally with its own bar) as something of a social club. But it is important not to confuse those who use the club atmosphere for their own social pleasure and those who use it as a medium for the lobbying of their peers. While the former may lack a certain commitment to their role as councillors, the latter most certainly do not and may be among the most effective of councillors. Invariably affable, the social councillors are generally ineffective but tolerated by colleagues because they can be relied upon to follow the party or majority line. In fact the absence of passionate idealism or the pursuit of power among social councillors can be viewed as a positive characteristic in terms of the personal and social relations between councillors. This aspect is often neglected in studies of local government. However as Newton commented, 'Council members lead a special social and political life which involves, to a greater or lesser extent, a separation from the wider society and a concentration of social interaction within the council' (1974). Social councillors are frequently an essential ingredient in both the social network and the party group system within a local authority.

In contrast to those who are involved in local politics for their own hard or soft gain, there are those whose involvement stems from less personal considerations. The classification which we have adopted below has close links with party affiliation and philosophy.

Amongst the gentry or pseudo-gentry Tories, the attitude was prevalent that as a member of a particular status group or section of society, it is one's duty to forward oneself for local government service. These *paternalists* (or *maternalists*) are most commonly found amongst the rural Tories. They feel that they have a clear planning mandate to 'protect' the environment, which in their terms represents a no-growth and no-change stance. They therefore try, as far as possible, to resist or restrict development in their area. Theirs is a truly conservative viewpoint.

There is an approximate parallel paternalist approach on the Labour side (usually in the form of the traditional working-class trade unionist), but this was not evident among the Labour planning councillors interviewed. The more common motivation derived from socialist principles was that of the *activist* and *agitator*; councillors who constantly challenge the status quo. The Labour councillors are urban-based and have taken a particular interest in the proliferation of office development. While they support growth, they seek employment opportunities for blue-collar and unskilled workers and have objected to the speculative nature of many of the office development proposals. Residential development and non-office, employment-generating proposals have been welcomed. But where office developments have been permitted in Reading, despite Labour opposition, considerable effort has been made to influence the design and appearance of the schemes. There was little evidence of a 'jobs at any price' attitude.

Get-involved councillors were ubiquitous. They were well exemplified by, but by no means confined to, the community-politicking Liberals. To get involved was the single most cited reason for becoming a councillor. It is a socially acceptable response which indicates neither too strong an element of self-righteousness nor any degree of vested interest or personal benefit. Often these councillors have been very active in their local communities and may have risen through the ranks of residents' groups, local voluntary groups and parish councils. While many remained in these organizations once in office, a few severed direct links for fear of being seen in too close a liaison with specific pressure groups.

The impetus to get involved often derived from specific events or issues. Frequently, the claim of being a reluctant councillor was aired but such claims should be treated with scepticism. Doubtless a few were genuine ('I was asked to stand on four separate occasions'), but many through their position in the community could afford the luxury of not being seen to seek office.

The actual events which encouraged people to stand as candidates were varied. One Tory councillor was initially approached by an alderman who told him 'business people like yourself should be involved in the running of the borough'. He was further persuaded to stand by the knowledge that a swimming pool was to be built – as an ardent swimmer he considered that through his involvement he could ensure that a competitive (rather than a social) pool would be built. Others in rural areas were sparked into action in response to perceived threats to their villages. Action was demanded by residents and those individuals who were already active in their communities were obvious candidates to stand for election.

Finally, mention needs to be made of the *professional politician* who sees local government office as one of the first steps in the career ladder to Parliament. The success rate of such councillors is undoubtedly low; there are, after all, only 650 seats in the House of Commons but 332 district councils, most with at least thirty councillors. While parish councillors often become district and/or county councillors, a local government position is by no means a prerequisite of a parliamentary seat.

Roles of planning councillors

While there are often close links between the motivation of councillors and their role in council, a motivation does not necessitate a role. A cross-tabulation may be envisaged which highlights likely and unlikely combinations of motivations and roles (see Table 4.4). Three major planning roles may be identified.

First, there are what Newton (1976) termed the 'policy brokers'. We prefer to label them 'policy administrators'. They are the councillors who come to grips with the complexities of policies. A distinction is worth drawing between councillors who substantially affect policy (the 'policy directors') and those (the 'policy upholders') whose activities are largely in support of existing policies. The latter, by virtue of their influential positions and knowledge of the limits of planning, often adopt the role of restraining the committee from taking 'non-planning'-based decisions. Of course individual policy administrators may play both roles to a greater or lesser extent. The principal policy administrators tend

TABLE 4.4: Likely and unlikely combinations of motivations and roles: survey assessments in brackets

		Policy administrator	*Place agent*	*Council filler*	
Hard gain	Vested interest	✓	✓	X	0
	Professional politician	✓(2)	X	O	2
Soft gain	Para-careerists	O(1)	✓(1)	O(1)	3
	Status-ticians	O	O	✓(1)	1
	Social councillors	X	O	✓(1)	1
Paternalists		✓(2)	✓	X	2
Activists		✓(1)	O	X	1
Get-involved		✓(4)	✓(3)	O(2)	9
		10	4	5	19

✓= likely O = possible X = unlikely

Block 9 A policy administrator

Mrs X was chairwoman of a planning committee in a district of Central Berkshire for three years. She had been a district councillor for four years before being given the chair of planning.

She is motivated by a belief in the need to get involved, to make a contribution to the locality. She is a political animal with a strong belief in the party ticket at local elections – 'people know where you stand'. Her Tory stance is one of protecting individual rights, keeping rates low and helping industry thrive.

She sees the job of planning to keep a balance between the needs of the development industry and local residents. Representing a more rural ward she is keenly aware of the need to protect villages from 'overdevelopment'. She is strongly opposed to village coalescence and physical incorporation into large towns because the 'village spirit is spoiled with increasing size'.

As chairwoman she has cultivated warm and strong links with her officers. She meets them regularly and is a firm believer in being loyal to them. 'They have a hard time with the amateurs on the committee.' The main job of the committee, as she sees it, is to stick to the policies – the less the deviation the less the danger of setting unwelcome precedents. She is an upholder of the planning system and tries to keep discussions in committee to planning considerations.

In committee she is efficient and well informed, keeping the meeting moving on quickly, but diplomatic and tactful with more naive members. Her manner is partly responsible for the more relaxed, more informal atmosphere of 'her' committee in contrast to others in Central Berkshire.

to be restricted to fairly high calibre, committed and ambitious councillors, often drawn from the paternalists, activists, and perhaps even those with vested interests. Block 9 presents a potted description of one policy administrator, who might be classified as a maternalist and a policy upholder.

Virtually by definition, councillors who wield very considerable power in policy formulation are few but many more can be involved in minor policy shifts. The major planning policy directors were generally to be found in positions of influence (chairpersons or party spokespersons) on the planning committees. Obviously, their policies have to accommodate the interests of other committees but there were few indications of undue pressure from non-planning councillors. Two exceptions, however, are worthy of note. First, at Berkshire County Council and particularly over the 'Heseltown' issue, the Conservative party leader took on an extremely active and influential *ex officio* role on the planning committee. Second, in Bracknell District Council after the 1983 elections and after the Heseltown allocation, the Conservative party leader began to take an active part in planning matters.

It might appear from Table 4.4 that we have been too generous in attributing the status of policy administrator to a large number of councillors. We interviewed the chairperson of all four committees and this boosts the proportion of policy administrators. None the less the general consensus is that a number of councillors make policy contributions, although their translation into outcomes obviously depends upon the distribution of power within a committee.

A role easier to play than the policy administrator is that of 'place agent'. The place agent is a parallel to the 'people's agent' identified by Newton (1976) and relates closely to his 'parochial' motivation for council service. The people's agent was concerned with representing the problems of individuals on a ward or district-wide basis. It is not often possible to perform this precise role on the planning committee, although there was one Liberal councillor who regularly represented the interests of disabled people at the planning committee (facilities and access to buildings).

The place agent, however, is very much a planning phenomenon; many councillors consider it to be their prime duty to resist unwanted development in their ward. The role of the place agent is firmly entrenched in many local authority planning procedures: discussion of an item usually begins by asking (formally or implicitly) the local ward representatives to make their comments. Should no comment be forthcoming, it is unlikely that other members will prolong consideration of an item. Not surprisingly, one of the main complaints of planning officers and policy administrators is that rank-and-file councillors are much too parochial in their attitudes and fail to consider the interests of the district as a whole.

To a greater or lesser extent, all planning councillors take on the role of place agent, although it is the paternalists and particularly the get-involved councillors who develop the role most fully. Effectiveness varies but most attributed any success they achieved to obtaining a very firm grip of 'the facts' and to having close

Block 10 A rural Liberal councillor

This councillor epitomizes three aspects of our Central Berkshire findings: the place agent, the anti-development attitude, and Liberal community-based politics. As the headteacher of the village primary school and district councillor for nine years, he was well respected within the community – a man who served the area well. His move into politics began at parish level, and he was reluctant to stand for the district, but local Liberals eventually convinced him that he was the 'right man for the job'.

He was politically more interested in local rather than national issues, but well aware of the political element within the planning authority. With a Conservative majority he would never obtain a committee post despite suitable experience, but had served as the Liberal planning spokesman for a period. Within the committee an influential member, realizing the importance of being well informed and consulting other members in advance, both to seek support for his own proposals and to obtain other views on non-ward items.

He was a conscientious councillor, with an estimated thirty-six hours spent on council work during the calendar month prior to the interview (equivalent to two or three evenings per week). Planning-related activities, including the time spent on personal and official site visits, consultations, reading reports and travelling to meetings, involved nineteen hours' work. The rural nature of the ward means only four to five applications per month; with only two or three of any significance. Investigation of applications included telephone calls to the district officers (one or two per week), attendance at parish council planning subcommittees if requested, consultation with any local residents likely to be affected by a proposal, and in some cases discussion with applicants to seek modification of submitted plans.

In planning terms he was very much the place agent, defending the settlement and its people against development. Because of the development threat to the local village the planning committee was regarded by him as the most important one. A man of strong principle, he was fiercely opposed to the subjugation of local planning authorities by Whitehall. He was the only councillor to vote against an application prompted by the Heseltown allocation within the district, and within the planning committee expressed respect for those members who 'stuck to their guns' and were consistent in their views.

This councillor was a genuine defender of place and community, espousing the need for local, not central control, and attempting to stem the rising tide of development within his area.

links with their constituents. Block 10 describes a Liberal place agent.

Finally, there are the 'council fillers'. They are the fodder for the political machine and have negligible impact upon policy or even minor decisions. Faithfully toeing the party line, they are agents to be manipulated by the more ambitious (see Block 11). They are generally naive of planning considerations and through a lack of information often fail even to obtain the decisions they wish within their own wards. Said one infrequent contributor to chamber discussion, 'I don't agree with a lot of the decisions we take but I'm an easy-going person and I won't speak out against the majority view.' Two distinct categories of council filler can be identified. Some can be classed as committee lightweights, ready

Block 11 An urban Conservative councillor

'I am easy-going about things – I think our planning policies are often too restrictive, but I generally go along with the rest of the committee.'

A councillor with the attributes of a council filler who was none the less the chairman of an important committee. As such, however, he could be considered as a front man who gave a presentable image of the Conservative party but who was easily manipulated or bypassed by other party members.

As a middle-aged small businessman with no real political ambitions, he was asked to stand as a Conservative candidate in the town in the late 1950s. He was already interested in local affairs, and the party was encouraging businessmen to serve on the council. This request coincided with discussions about a sports facility for the town, and being a keen athlete, he was eager to ensure that it would cater for competitive as well as casual sporting activities. With reservations he agreed to stand, and so began a sustained involvement with local politics at both district and county level.

The initial motivations for service suggest a get-involved categorization, but by the 1980s a para-careerist was a more apposite description. His business commitments were sufficiently flexible to permit a number of years of simultaneous service on both the county and district councils.

Unlike so many of the rural Conservatives, he was by no means a place agent. As an urban Conservative he had paid his political dues by standing in wards where electoral success was unlikely. Since local government reorganization, however, he had been in office continuously.

In planning terms he was very much a responsive rather than an initiating councillor; taking notice of objectors and respecting the views of the officers, but rarely investigating applications or consulting officers or other members before a committee meeting. His main role on the committee was to vote with the party line when necessary. The quotation concisely sums up his attitudes to planning and local government. In every sense a party man.

to venture an opinion and often competent in constituency casework, but simply lacking influence and authority at the planning committee. The second group are aptly described as committee deadweights; these councillors make very few attempts to influence planning outcomes (and these almost exclusively within their own ward) and have little involvement with committee work or discussion. The deadweights are to be found mainly amongst the status-ticians and the social councillors; the lightweights may include get-involved councillors.

The council filler role is the one most affected by length of service. Without prior experience of planning most new planning councillors need six to nine months to grasp the basics of the system, and one to two years before feeling confident in committee discussion. A number of the deadweights on the other hand are older councillors and have given many years of service to the council. They may have been more active in policy matters or local affairs in the past, but have settled into a routine of 'minimal involvement' with council business. A safe ward is usually characteristic of such representatives.

The pressures on planning councillors

The impact that elected members have on the planning system is not simply a function of their individual motivations. There are a whole series of influences which act upon them in their role as planning councillors, a number of which will be examined in this section. Greater detail, and the additional influences of public accountability and the political decision making process, are considered in Witt and Fleming (1984).

One of the questions in the district councillor survey was concerned with the degree of importance attached to other agents. The replies were coded on a five-point scale from 'very important' (scoring 5) to 'of no importance' (scoring 1). The results are shown in Table 4.5. It needs to be emphasized that their values represent general levels of influence and that pressure from a particular source can override seemingly more important influences on any given application. For example, despite the obvious importance of officer recommendations shown in the table, all the councillors indicated that they would not hesitate to overturn the recommendation where necessary.

Officer pressure

The most important influence upon overall development control decision making is undoubtedly the advice of the planning officers.

TABLE 4.5: Proportion of councillors attributing high and low levels of importance to various influences acting upon them (in percentages; sample size varies between influences: 15–19)

	% Important or Very important	*% Little or No importance*
Officers' recommendation	74	5
Community group representations	42	11
Parish council representations	40	20
Committee majority views	39	22
Party group views	35	41
DOE advice (circulars)	28	50
Individuals' representations	22	11
Applicants' representations	21	42
Chairpersons' views	14	64

Since a local planning authority has to operate a process which is fairly tightly constrained by the legal framework and the views of central government, planning councillors are inevitably dependent upon their officers for advice on their freedom of manoeuvre, and the likely consequences of their decisions. This is not to suggest either that councillors have no influence upon the local planning process or that they accept the officers' views on all occasions. It is simply the fact that the officers are an integral part of the planning system, and, as further sections will indicate, are an ever-present factor in councillor deliberations. Because of their position, it is not appropriate to examine the influence of the officers as a single, discrete element. Their influence will become apparent during the consideration of the development control process in operation.

Party politics

Only since 1974 has party politics emerged as a highly potent force within a local government in the shire counties such as Berkshire. Prior to that date it was restricted to the main urban areas. Reorganization heralded a rapid decline in the importance of the 'independent', and a rise in party political consciousness at both county and district level. While the rating issue, for example, has become almost a classic in its demonstration of party differences, planning has had a relatively low level of party politicization, and even the County Planning Officer noted how little priority was given to strategic planning by county councillors during the period of structure plan preparation (Stoddart, 1983).

Party political stances are most strongly evident in Reading Borough Council, and frequently reach the planning arena. A local commentator caricatured the party viewpoints thus: 'the Tories would say it [an application] was horrible but they ought to approve it, the Labour members would say it was capitalist greed and ought to be rejected, and the Liberals would seek better access for the disabled' (*Red Rag*, October 1983). In Reading, there is more than a grain of truth in this description, but it also highlights some of the party differences within Central Berkshire as a whole. The Conservatives have generally favoured commercial and industrial development, and have been much more pro-office development than the Labour party. The latter has consistently opposed rapid office growth, particularly of speculative developments, and bemoaned the declining public sector housebuilding rate in Bracknell and Reading. Most politicians from rural areas have objected strongly to any significant development in the countryside, and it has been the Conservatives and Liberals leading the fight against the continuing high rates of residential growth.

Ideological differences are also evident at the level of individual applications. Labour councillors often sought to impose conditions upon planning consents which the Conservatives considered as unacceptable restrictions of enterprise and initiative. The Liberals swayed back and forth across the middle ground, although they supported neighbour and neighbourhood objections to a greater extent than the other groups. While planning criteria often submerged the party affiliations, it was significant to note that when contested votes occurred in Reading, the committee almost invariably split along party lines. During eighteen committee meetings the Labour group never broke ranks, and only one independently minded member voted against the Conservative group position. Within the other districts the extent of party politicization in planning was much lower, and its manifestation at committee was less overt.

Ward pressure

Pressure from the ward is an important influence upon most committee members. It can emanate from a number of sources: individual objectors, influential individuals or organizations within the area, the parish council (except in Reading BC), local community groups and political opponents within the ward.

INDIVIDUALS

Since councillors operate on a ward basis, there is considerable pressure upon them to serve the interests of both the constituency as a whole and of individual constituents. Most people who contact their local councillor on planning matters do so in order to object, usually to a recently submitted application or to unauthorized development or use of land.

It is a point often overlooked by planning officers and developers that councillors frequently encounter anti-development feeling through personal contact (visits or telephone calls), whereas formal objections to the planning authority and the occasional contacts between objectors and developers are in the more impersonal form of letters. The planning officers can withdraw into their bureaucratic shelter if heavily criticized by a member of the public ('we are only carrying out our statutory duty in accordance with current government policy'). Councillors however are more vulnerable to charges of not serving the interests of their constituency. There is also the trite but valid point that ward pressure on councillors is sustained by the likelihood of encounters with dissatisfied constituents during day-to-day activities within the area.

Place agent councillors, who consider themselves to be closely involved with the affairs of their area, are most likely to spend time and effort in investigating a proposal. Such councillors often seek the views of residents who might be affected by the proposal, and unless personally in favour of the scheme (unlikely if it is 'harmful' enough to provoke local opposition) will assure them that (s)he will resist the application at committee. Other councillors are less inclined to investigate every application which provokes opposition within the ward. Many minor planning applications will incur an objection from the nearest neighbour and councillors are reluctant to waste scarce time dealing with what they see as trivial matters. On more important applications however the status of the councillor within the community, and on occasion, future electoral prospects, may be dependent on the extent to which local objections are supported (see Block 12 for an example of a councillor taking considerable trouble to demonstrate concern and support for local opposition to a planning application).

PARISH COUNCILS

A formal source of pressure within the ward is the parish or town council (except in Reading BC which does not have local

Block 12 Pacifying ward pressure

Pressure upon a councillor from the ward is a general feature within local government, but has particular importance in electorally marginal wards. Members who were elected 'by the skin of their teeth' are forced to pay serious attention to their political image and standing within the constituency, and will take every opportunity to publicly promote or defend the local interests.

One Central Berkshire councillor in a marginal seat provided a useful example of the interaction of local politics and the development control process, with the intention of pacifying potential ward opposition. This Conservative councillor had been newly elected in May 1983 with a surprise victory in a traditionally Liberal ward; a case of a paper candidate serving his political apprenticeship in a difficult seat, who won because of the large swing to the Conservatives. Once elected, however, he intended to remain in office. An opportunity arose to bolster local support with a planning application for a small block of flats.

This proposal aroused strong opposition in the neighbourhood, provoking local press interest and a number of written objections to the district planning authority. The councillor, not a planning member but with an interest in the subject, spent considerable time meeting with a number of the objectors and discussing their views. Assurances were given of his intention to resist the proposal at the planning committee.

At the meeting, however, he adopted a slightly different course of action. Instead of agreeing with the officer's recommendation of refusal and making a constituency speech attacking the application, he requested a site meeting. The site meeting was intended to pacify the local objectors, particularly the residents in the affluent area who would have been most affected by the scheme. The visit of the planning committee to the site, and the opportunity for the objectors to put their own case to the councillors, would demonstrate the concern of the local member and his effectiveness in serving the community.

The councillor readily admitted that the underlying reason for this close involvement, and particularly for the site meeting, was to avoid any Independent candidate from gaining electoral support at the next election. In a marginal seat any split within the Tory vote could easily allow the Liberals to regain power. We suspect that this sort of motivation is more widespread than is commonly recognized.

councils). These councils have a statutory right to be informed of planning applications within their area, and they forward their written comments directly to the district council officers. As such they represent both a tier of local government and a pressure group.

The principle of parish (or town) council involvement in planning was welcomed by all of the interviewees but criticisms and reservations were widespread. The main benefits were seen to be their local knowledge and representation of local feeling at the grassroots level. Criticism, expressed by eleven of the fifteen (non-Reading) councillors and supported by opinions expressed during public committee meetings and informal discussions, seemed to reflect five common issues: political tension, personality clashes, parochialism, planning naivety and apathy.

The relationship of a parish council to the local district councillor can range from being a close ally to a highly critical opposition body. This was reflected in the questionnaire results, with six of the councillors claiming a good or very good relationship, five only fair, and four a poor or bad one.

In some cases there is considerable cooperation, particularly where the district councillor also serves at parish level. A planning committee member in such a position is well placed to advise the parish council when it discusses planning applications and submits its comments to the district council. A united front of local planning councillor and parish council is a combination not lightly ignored by other committee members.

At the other extreme from dual membership is the situation of little contact, for either personal or political reasons. This can lead to opposing views being forwarded to the planning committee from the two official channels of local representation. In these situations the view of local planning members invariably prevails over the parish council comments.

In between the two extremes some councillors use their parish council(s) as a watchdog, scrutinizing all the planning applications for the area and bringing to their attention those which cause concern.

COMMUNITY GROUPS

In terms of their pressure upon councillors, local community or residents' groups fall mid-way between individual objectors and parish councils. Because they form a group, with a structure and a title, members of such organizations gain certain legitimacy as a voice of their area. However a community or resident group is not a formal part of the development control system in the same manner as a parish council. The closest structural relationship between local groups and a district council in Central Berkshire occurs within Wokingham DC, where consultative committees exist for the two major developments of Lower Earley and Woodley Airfield. Residents' groups within or covering part of each development, together with the local town council and representatives of the district planning committee, consider all applications relating to their area and forward their views to the district.

Councillors have rather ambivalent attitudes towards community groups. Most recognize that local groups do represent certain grassroots feeling, and have an important role in bringing issues to the attention of councillors and the local public. If the relationship between councillor and group is a close one, then it is possible for information to be disseminated to local people through the community group, and for considerable cooperation to occur in the cause of achieving the aims of local residents. The relationship of one urban Labour councillor with the local group was very close, to the extent that a political opponent labelled the group as 'the . . . Labour party under a different name'. A rural Conservative indicated a rather different role towards the local association: 'I often act as an information service, and soften the blow where necessary by explaining that outline consent has already been granted but that I will fight to get a better deal over the details.'

Other councillors are slightly hostile to local groups, because they create tension and conflict within the area. Because of the localized nature of such groups, and the fact that they are principally lobbying rather than decision making organizations, they are invariably parochial in their attitudes. Councillors may sometimes be outflanked by a local group and forced into a defence of actions implemented at district or higher levels, which appear to work against the best interests of the area.

It is perhaps surprising that community groups, often caricatured as elitist, unrepresentative and belligerent, ranked higher than parish councils in Table 4.5 (see p. 133). But of the eighteen councillors who had at least some contact with local groups, ten described the relationship as good or very good, six as 'mixed' and only two as poor.

APPLICANTS

In the same way that objectors and community groups inform councillors of their views, some applicants attempt to influence the committee's decision by direct lobbying. This often occurs when the applicant senses from discussions with the officers that a negative recommendation will be presented to the committee. It can also happen if an applicant who has a good understanding of the planning process realizes that the proposal is likely to achieve a more sympathetic response on political or local interest grounds than on 'pure' planning principles or policies.

Pressure from applicants or their agents is much less widespread than that from objectors. Of the nineteen councillors in our sample, seven were contacted by applicants less than once per year. Applicants are likely to contact a councillor in two situations. First it is the local member whose support is sought. This includes many of the 'desperation' cases where the officers are intending to recommend a refusal and the applicant switches to the committee for support. As with objectors, most applicants in this situation turn automatically to the local member as the most obvious source of potential backing. Second, certain shrewd agents or applicants will engage in selective lobbying of the more influential members. They realize that on many occasions the views of these councillors are decisive in influencing the outcome of committee debate.

Pressure on councillors is not restricted to letters, either to individual members or to the whole committee. Especially with local applicants (usually individuals or small firms), personal lobbying may be carried out through chance or deliberate contact within the ward. This form of pressure is often not appreciated by members, who may resent being lobbied at work, in the street, or when relaxing in the local pub.

On rare occasions, but invariably concerning large developments, members of the planning committee are invited to inspect the site with the applicants and agents, or visit another development by the same company.

A further means of pressure upon a committee by an applicant was the suggestion of gain to the local community from a development. At one end of the spectrum, applicants would generally convey a positive, flexible and helpful attitude to the committee, while at the other extreme were cases of strong applicant pressure upon the council. A blatant example of 'moral blackmail' is outlined in Block 13.

The system in operation

In this chapter we are primarily concerned with the role of planning authorities, particularly the elected members, in meeting the range of pressures related to speculative housebuilding. Having examined the planning councillors of Central Berkshire, we now turn to the operation of the development control system, which is crudely pictured in Figure 4.1. Each authority will have its own variations, but they all share a basic pattern. The full cycle from submitted application to decision may take three to eight weeks for typical applications but up to two years for complex proposals requiring legal agreements.

Block 13 An example of applicant pressure

Applicants can apply pressure upon planning councillors in two ways: by stressing their readiness to go to appeal if permission is refused, or by offering community gain in exchange for permission. An extreme case of the latter was perfectly illustrated by two linked applications submitted to Wokingham DC in 1982.

One application related to the site of a scrapyard in a residential area which had caused disturbance to local residents for fifteen years, despite attempts to control and re-locate the business by the council. Local councillors continued to receive complaints and demands for action, and therefore welcomed with open arms the proposal to redevelop the site with five town houses and a bungalow. There was of course a price to pay; the other application proposed an industrial/storage development together with ancillary offices on a site (owned by the same applicant) within the narrow green strip between Reading and Wokingham. Through the officers the applicant requested the committee to consider both applications together and stated his readiness 'to relinquish all established rights at the scrapyard if the district council for their part granted permission for the development of both sites in the manner proposed'.

The two items went directly to the full committee, accompanied by written reports from the officers and their recommendations of consent in both cases. Although the industrial proposal was just outside the limits of a built-up area, and 'would not under normal circumstances be entertained under policy restraints for the area', it was emphasized that 'the proposals would achieve considerable planning gain by the removal of the non-conforming scrapyard'. The officers had agreed upon a draft legal agreement with the applicant which would prevent any other applications on either site, prevent the commencement of development until the scrapyard was cleared, and remove the existing use rights for the scrapyard site.

The committee was deeply divided over the issue. The councillors representing wards near the scrapyard were strongly in favour; this was 'a golden opportunity for local residents' in the words of the ward member. The improvement in the state of the other site (itself something of an eyesore) was cited as a supporting reason, together with the area of land to be given over to the public. Planning gain would result from both applications.

Opposition came from two quarters. The committee members for wards near the greenfield site denounced the selfishness of some of their committee colleagues; planning gain in one area was planning loss in their own. Objections were marshalled on a number of grounds: policy presumptions against development; traffic flows; preservation of the Green Wedge, and precedent (the site formed one corner of a roundabout, and permission would open up the other three sites for development). Opposition was also voiced by a number of influential members who were generally policy administrators. They opposed the consideration of both applications jointly; each should be considered on its merits. Planning policies were stressed, together with fears over precedent and the consistency of committee decisions. As one councillor commented, 'At times this committee doesn't look like a planning committee; here we have emotional arm twisting.' Towards the end of discussion the officers were forced to defend the linkage of the two items, but despite comments about the morality of the situation one vote was taken for both applications. The outcome was a convincing fifteen to four majority for the officers' recommendation of approval.

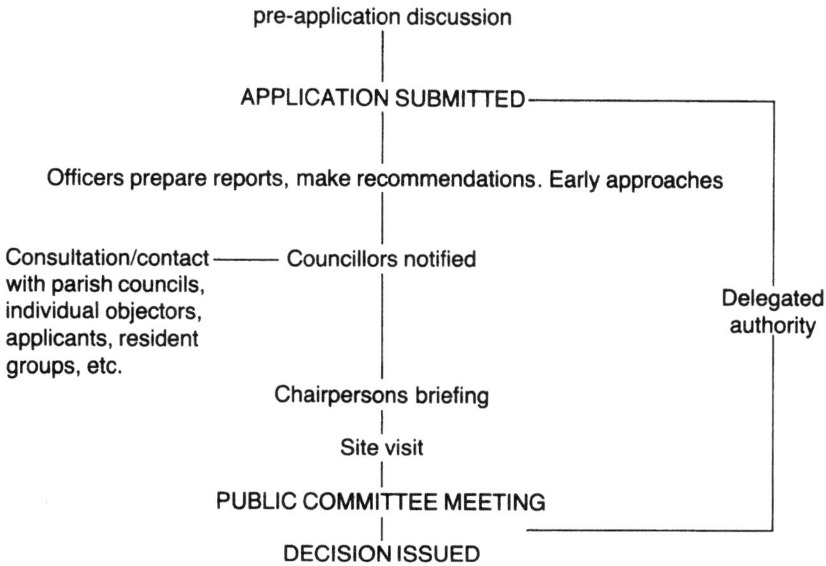

FIGURE 4.1: *Simplified outline of the development control process*
Note: The elements in block capital letters indicate the public face of planning, the others are the private face of planning.

Delegation agreements

At an early stage within the development control process, applications are divided into two groups: those for committee and those for officer determination. The large numbers of applications within the area, and the trivial nature of so many of them, encourages members to delegate minor decisions to their officers as empowered by the 1972 Local Government Act, Section 101.

A common theme runs explicitly or implicitly through all the delegation agreements: if a written objection is received, the item will go to committee. Thus the members can be seen to be discussing every item to which their electorate has articulated concern. As Table 4.6 indicates, the officers determined the highest proportion of applications in Bracknell DC, where even multiple dwelling applications which receive no objection can be delegated. In the other three authorities virtually all single dwelling applications are committee items.

Delegation agreements are flexible and can be renegotiated periodically. Since the uncertainties of local government re-organization when both members and officers were feeling their way,

TABLE 4.6: Delegated items during three cycles of the four Central Berkshire planning authorities

	No. of new dwelling applications		*Total no. of applications*		
	Total	*Delegated*	*Total*	*Delegated*	*Percentage*
Newbury DC	37	1	152	73	48
Bracknell DC	28	18	157	113	72
Reading BC	25	0	254	102	40
Wokingham DC	76	0	433	121	28

there seems to have been a general strengthening of officers' delegated powers. This may reflect the growing trust of the committee and a decline in their early zeal to be involved in the making of every decision. Only in Wokingham DC has there been a recent renegotiation of delegated powers. In fact in certain authorities, most of the interviewees were not aware of the precise details of the delegation agreement.

The more flexible agreements gave officers considerable discretion about which decisions could be made without reference to their committee. But even where the agreement clearly set out categories of applications for committee and officer determination, the officers did not always exercise their full delegation rights. Potential delegated applications in sensitive landscape or conservation areas, or on sites with a history of contentious proposals, were generally sent for committee consideration. The senior officers were aware that a policy of jealously operating up to the limits of their powers would be likely to provoke resentment from councillors, and so strain working relationships with the committee.

It is our impression that place agents were most anxious about losing their decision making powers through delegation. They feared a loss of their role if too many local decisions were taken out of their hands. In contrast, policy administrators tended to see things in a broader perspective and had glimpsed the inevitability of many of the minor decisions being taken by officers.

Early approaches

As we noted in the last chapter applications are rarely presented to the local planning authorities without some prior discussion. These discussions are usually held with the officers although members may be contacted by local applicants seeking general advice or specific information prior to submission.

After an application has been submitted members may attempt to modify the application and/or influence the officers' response before a formal recommendation has been made. Such action requires contact with the officers at an early stage, often weeks before the committee meeting. The committee chairs have more opportunity to influence proceedings than backbench councillors and often have slightly wider motives for such action. Backbenchers usually restrict their prior approaches to ward items, with the intention in most cases of obtaining changes in the proposal or conditions from the officers. The chairperson, primarily in those authorities where the dominance of the councillors is stressed, seeks to impose his/her will on the planning officers and achieve greater control over the course of planning within the area. Early approaches by the chairperson can be an important element in the power struggle with the Chief Planning Officer over the effective locus of control and decision making within the authority.

The written report

In Reading BC, Bracknell DC and Newbury DC written reports were prepared for all development control items to be determined by the committee. In Wokingham DC however they played a less important role; up until mid-1984, verbal reports were presented to a subcommittee, with written reports prepared only for major proposals and the authority's own applications.

The written reports were distributed to members seven to ten days prior to a committee meeting presenting the salient details of applications to councillors: the site, planning history, quoted or edited responses of consultees and objectors, the policy context, and other aspects which appeared relevant to the officers. Finally the officer's recommendation was presented in sufficient detail that it could be directly implemented as the decision.

In each district written reports were distributed to all members of the council; but did even the members of the planning committee read them thoroughly? All the chairpersons claimed to read all of the contents and we have little doubt of this, since they required the information for the management of the meetings and also had agenda briefings. Amongst the rank-and-file councillors, all of the Newbury and Reading interviewees also claimed to read every report. In Bracknell, however, rank-and-file councillors admitted that their reading was limited to reports concerning their wards, plus perhaps a few others of interest. It is difficult to tell whether the Bracknell councillors were merely more honest, but we think not. It seems likely that particular committees can build up a traditional approach to such matters as report reading. Said one of the influential Bracknell members,

> 'I get the planning reports seven days in advance, along with many other reports. Often I simply don't have the time to read them in any detail, so I flick through them checking application type against recommendation. I only select those recommendations with which I think I'll disagree to read in detail. And I'm sure you've seen me at committee reading the reports as we go along.'

Personal investigations

In addition to reading the officers' reports, there were a variety of ways in which councillors gathered information and views in their preparations for a committee meeting. Visiting the site enabled the councillor to gain a better impression of how a proposal would relate to its surroundings. Particularly during the early years of service on a planning committee, councillors often found it difficult to visualize a proposed development and its likely impact merely from site maps and plans. Personal site visits within the ward were expected of all planning councillors, and were virtually prerequisites for proposed committee site meetings.

Informal consultation with the officers was a second common form of preparation. Where the publicization of an application aroused local concern, a councillor required clarification and explanation from the officers, often to inform (or correct) objectors. The likely recommendation of the officers and the progress of any discussions with the applicant was also sought. In some cases a councillor awaited the officers' views before deciding upon an appropriate stance to adopt in relation to local objectors and perhaps the parish council. A balance is required between outright support for the local case and retaining credibility in the eyes of officers and fellow councillors. Discussions with officers were either conducted by telephone or combined with a visit to the planning department in order to examine the plans and drawings.

Over half the interviewees consulted with other members of the committee. This was done for various reasons: to discover local reactions to applications outside a member's own area; to enlist support from other councillors for a particular decision; or simply to obtain other views about applications prior to the meeting. One member of the Bracknell committee admitted the existence of a general system of 'horse trading' among the rural (Conservative) members of the district, with a tacit agreement that they would support each others' views on particular applications within their wards. There is an implicit assumption amongst many planning councillors that the local member knows best, especially for the smaller-scale proposals.

The recommendation

The planning officer's recommendation is the key component in the decision making process. Whether presented orally or in documentary fashion, the information officers give to committee almost invariably contains a recommendation. It is often naively considered that officers provide professional and technical information and advice for the members who then use their political judgment in evaluating the information and issuing a decision. But this is too simple a model for any field of local government responsibility, and for planning in particular. The distinction between technical and political considerations is unclear and the force with which a recommendation is made and supported at committee is a very important variable in influencing decision.

The separation of recommendation and decision is both a weakness and a strength of planning authorities. The weakness derives from the fact that applicants can exploit differences in the views of officers and members. As we will show, it is more common for Central Berkshire committees to reject recommendations for approval than for refusal: committees tend to 'block' rather than 'facilitate' applications. In taking blocked applications to appeal, an applicant can, through statement and cross-examination, highlight the difference in views and thereby strengthen his/her case. Facilitated applications often indicate strong lobbying by the applicant to bypass the resistance of the officers. The strength lies in the bargaining counter given to the officers; they can agree in principle to a proposal but the applicant cannot be certain of gaining consent. Further concessions may be obtained by indicating that the committee will take a firmer line.

The separation also emphasizes the distinct roles of officers and local politicians. The officers strongly believed that their duty was to advise, and proffer a recommendation, on the basis of planning grounds (e.g. policies, guidelines and professional wisdom) only.

Pre-committee meetings

Before the planning committee meets to make decisions, there are a variety of private and informal meetings. These present opportunities for discussion and the exercise of influence prior to the public committee. In the course of our research we noted five types of pre-committee meeting; two of them common to all the districts – the *chairperson's briefing* and the *committee site visit* – while three were unique to one district – the *planning committee party group meeting* (Reading), the *management group* (Reading) and the *committee working party* (Newbury). Here we will only be concerned with the first three.

CHAIRPERSON'S BRIEFING

One of the 'private faces of planning' closely linked to the public committee is the chairperson's briefing. This is held for the benefit of the chair and vice-chair one or two days prior to the planning committee meeting, and takes the form of a run-through of the agenda by the officers.

This interaction can be viewed at two levels. Considering it first purely in terms of local government planning procedures, it is intended to familiarize the chairperson and vice-chairperson with the items on the agenda and the officers' views and recommendations. This aids the efficient management of the committee meeting and allows the chairperson to field many of the queries from the committee without referring everything directly to the officers. For reasons of personal pride and 'job satisfaction', a chairperson will wish to conduct the committee meetings in an efficient manner, and a full briefing by the officers is a key preparation. The briefing enables the two senior councillors to clarify uncertain points, discuss particular details and to receive advice and assessments from the officers which would not be ventured at the public meeting.

The briefing is also a private although highly structured setting for the interplay of political and professional pressures. It provides an opportunity for the senior councillors to covertly influence planning within the area. The agenda and recommendations have already been circulated by the time of the briefing, and the officers are unlikely to reverse many of the recommendations. But especially with briefings held a day or two before the meeting, there is time for councillor suggestions, modifications or criticisms to be acted upon. There are occasions when a councillor brings fresh information or hitherto unarticulated local concern to the briefing, which affects the officers' recommendation.

The officers also gain an indication from the councillors of their reaction to each application and recommendation. They would therefore be aware of the views of two influential members prior to the committee meeting.

The basic format of the briefing has been outlined but it is important to probe beneath the generalities. Although they are held in private, access was gained to two briefings.

The main impression from a briefing in a non-politicized authority was of subtle officer domination. Although the style and atmosphere were relaxed and informal, there was the sense that the briefing was given *by* the officers, *for* the benefit of the councillors. The officers worked through the agenda, elaborating or emphasizing points in their written reports, and showing the relevant plans. Many comments and queries were made by the two councillors, and one-third of the items were discussed in some detail. On a number of applications the officers had to defend and justify their recommendation, twice giving an assessment of the chances of appeal success which would not be repeated at committee. Opinions contrary to the officers' views were not withheld, and disagreement or doubts about the application were expressed on almost one-third of the applications. The chairperson signalled an intention to vote against one recommendation whilst the vice-chairperson indicated that he was firmly against four.

A constrast is provided by a briefing in a politicized authority. The roles of the two councillors in this briefing were strongly influenced by the party group system which operated within the council. As the planning spokesperson for the controlling group, the chairperson had primary responsibility for the party's position on planning policies and applications. The party group system, and the group's committee majority, usually facilitated the translation of the chair's views into decisions. At the briefing the officers were therefore justifying their recommendations to the key decision maker, and any modifications or reversal intentions of the chairperson was likely to be carried through at committee. For this authority the chairperson's briefing was the principal decision making arena.

COMMITTEE SITE VISIT

The second type of private meeting is the committee site visit, where place pressures are particularly evident. Its purpose is to assist the committee in their deliberations, especially where an appreciation of the actual site appears crucial to the decision.

The site visit is of great value to committee decision making, allowing finely balanced applications to be assessed in their actual setting. Time constraints however restrict the number of applications which can be deferred each cycle, and attendance at site meetings varies widely between districts, depending on the timing of the visits, the location of sites and the provision of transport by the council.

Despite its importance as a procedural device to assist with finely balanced decisions, the main significance of the site visit lies in its impact upon officer-committee relations. At committee much of the information involved in decision making is in abstract form, but when councillors stand in a field, or inside an empty building, the information is received directly from the environment. The officers' influence is considerably diminished. These factors are reflected in the high level of committee reversal of the recommendation when site meetings are involved. This will be taken up again later, but it is sufficient here to note that in Newbury and Reading, reversals at or following site visits were respectively eight and five times more frequent than for all applications at committee.

There are two components to these higher figures: the reduced influence of the officers on site and the fact that many of the deferred applications were known to involve officer-committee conflict. Since most councillors recognize the propensity for site meetings to lead to reversed recommendations, this procedure is regularly used by councillors attempting to swing a decision against the officers. The battle against the strict planning case is fought out on ground of the local councillor's own choosing, namely the actual site. The councillor can argue the case for reversal more convincingly than at committee, pointing out particular concerns of local residents, and encouraging other members to view the site from the most appropriate positions. A diligent local member, particularly a place agent, will often arrange for the committee to view an application site from a neighbouring garden or kitchen, and demonstrate with the aid of the plans the impact of the proposed development. Even if the attempt to prevent development fails (most site visit reversals result in committee refusals), the local councillor has done everything possible to influence the decision, and has answered the call of local protest or personal conviction.

The importance of local pressures at a site meeting is reinforced by the involvement of other interested parties within the development control process. A site visit is one occasion where, in some authorities at least, applicants, objectors, the parish council and other bodies can directly present their views to the committee. The applicant and/or agent, whether or not formally invited to speak, frequently attempts to justify the proposal and to dispel any concerns of the committee. The parish councils are often invited to send a representative, and in Newbury, any objectors who attend (or a representative if a group of them) are permitted to present their case. Site meetings are rarely advertised, and the presence of objectors is often an indication of the ward councillor either pulling out all of the stops to block a proposal or attempting to satisfy local protestors by enabling them to speak to the planning committee.

THE PARTY GROUP PRE-MEETING

One manifestation of the political element within local government is the party group caucus or pre-meeting. It has already

Block 14 District councillor activities and time commitments

In Britain, despite some periodic suggestions to the contrary, there are no salaried councillors in local government. Most councillors therefore have to fulfil their obligations in their spare time. So how much time do they devote to council affairs? The Maud Committee (1967) discovered that councillors spent an average of fifty-two hours per month on council matters. The Robinson Committee (1977), reporting ten years later, estimated a figure of seventy-nine hours per month. The increase was due to the effects of 1974 reorganization which reduced the number of councillors, and for many authorities brought in more party meetings.

In our survey we asked councillors to note the time spent on various activities during the last month. Given this looseness, the figures can only be approximations. Twelve of the nineteen councillors responded, two from Reading, three from Wokingham, three from Bracknell and four from Newbury.

Analysis of the results showed that the average time given over to council affairs was forty-four hours per month with a significant variation between the more urban and politicized authorities (Reading and Bracknell, fifty-five hours) and the more rural, or single-party authorities (Newbury East and Wokingham, thirty-five hours). A breakdown of this total time is given in Table 4.7.

TABLE 4.7: Breakdown of councillors' activities

Category	*Robinson Committee*		*Survey*	
	Hours	%	Hours	%
Attendance at meetings	23	29	21	48
Preparation	18	23	10	24
Travelling	8	10	5	11
Party meetings	5	6	3	6
Electors' problems	13	17	5	11
Other	12	15	–	–
Total	79	100	44	100

The figures are, of course, aggregates of varying individuals' commitments, with chairpersons generally having more commitments than backbenchers.

been noted that since reorganization in 1974 formal political structures and procedures have emerged within most local authorities. The factors of party discipline and collective responsibility are an integral part of local government in a politicized authority. As Green discovered within the controlling Labour group in Newcastle, 'Councillors often stressed that the pre-meetings should take the decisions and that everyone, including Chairmen, should be bound by these decisions' (1981, p. 133).

In all the Central Berkshire authorities, party pre-meetings were held prior to full council meetings. In Reading BC, the most politicized of the four districts, a similar procedure had been adopted for the Development Control Subcommittee meeting. These pre-meetings represent the forum for fixing the party position on the planning applications to be determined. Any discussion and conflict within the group is internalized, and the agreed position is then adopted by the whole group at committee.

The public committee

The public planning committee (or subcommittee) is the forum where most applications are formally determined by councillors. Only rarely do applications go to a higher committee or the full council. The committee decisions represent the outcome of the competing pressures which impinge upon planning councillors. Of particular importance is the tension between officers and councillors, reflected in the degree of correspondence between recommendations and decisions.

As outlined above, almost all applications due for committee consideration have been subject to formal or informal scrutiny and consideration by at least some councillors. Unless an application is deferred, the public committee represents the last opportunity for a councillor to influence a decision.

The public committee has other functions, however, in addition to decision taking. Discussion and determination of applications in open session is one component of public accountability within the local planning system. The meeting also acts as the formal stage for the expression of disagreement and conflict, both within the committee and between members and officers. Finally, the public meeting provides a platform for speech making by councillors, giving councillors the opportunity to make critical, supportive, or ideological statements intended more for wider public consumption than for their fellow committee members.

'The members' last stand': committee decisions and reactions

A committee's response to officers' recommendations can be classified into five main types: concurrence, blocking, facilitating, modifying and stalling.

Concurrence: with or without debate, the committee accepts the officers' recommendation and implements it as a decision.

Blocking: by not accepting an application for which officers have recommended approval, the committee refuses it and thereby blocks the application.

Facilitating: by approving an application for which the officers recommended refusal, the committee acts as a facilitating agent.

Modifying: in officer recommendations of approval, conditions are normally attached which the committee can alter, delete or supplement. In officer recommendations of refusal, the committee may alter the stated reasons for refusal of an application.

Stalling: the committee decides to hold a site meeting, or to defer an application for further negotiations or to obtain more information. This postpones the decision and often leads to a blocking, facilitating or modifying outcome at a subsequent meeting.

In the following discussion these modes of committee response will be explored more fully but first it is appropriate to indicate their frequency of occurrence. Table 4.8 presents the results of the survey of public committee meetings conducted in late 1982/early 1983. In Wokingham it was subcommittee recommendations which went to the full committee, and since the officers' recommendations were not accessible only three districts were examined. It is possible however to give an indication of the actions of Wokingham councillors by using data from one private subcommittee meeting observed in 1984; these figures are included in Table 4.8 for comparison. Both the data and the arguments from the three-district survey are augmented by reference to larger samples: work undertaken at Bracknell based on planning files covering the period 1974 to 1981 (Fleming and Short, 1984), and data from eighteen meetings of both the Reading and Newbury planning committees over the period late 1982 to mid-1984.

CONCURRENCE

In no authority was there less than a 60 per cent rate of total agreement between recommendation and committee decision. This of course relates only to items determined by committee; the overall extent of concurrence within an authority would have to include delegated items, over which there is implicit agreement.

TABLE 4.8: Proportion of applications (all and new dwellings) in each mode of committee response; three cycles of NDC and RBC, four of BDC in 1982/3; private subcommittee of WDC in 1984

		Items	%				
			Concurred	*Blocked*	*Facilitated*	*Modified*	*Stalled*
Newbury DC	All	79	75	3	–	11	11
	New dwellings	36	86	3	–	8	3
Bracknell DC	All	55	62	5	4	25	4
	New dwellings	15	60	13	–	20	7
Reading BC	All	152	70	11	–	10	9
	New dwellings	25	64	16	–	8	12
Wokingham DC	All	38	86	8	–	3	3
	New dwellings	10	70	20	–	–	10

By incorporating the figures for the proportion of applications determined by the officers, overall values for concurrence would rise to 87, 89 and 82 per cent for Newbury, Bracknell and Reading respectively.

In overall terms councillors influence only a limited proportion (11–18 per cent) of planning applications. *The development control process is primarily an officer-operated administrative system*, with many committee decisions little more than a rubber stamp for the sake of public accountability. The real test of the officer-councillor power relationship, however, concerns that subset of all applications which councillors *wish* to influence. Few councillors have an interest in many of the minor items. Therefore the 30–40 per cent of the more important applications influenced (in some respect) at committee indicates a significant, even if not overwhelming, impact upon the development control process.

BLOCKING

Blocking and facilitating decisions are outright reversals by the committee of a recommendation, with blocking more common than facilitation. Although the figures in the table are based on small samples, the longer-term data paint a similar picture. The Bracknell study gave a value of approximately 6 per cent for committee reversals, and the eighteen cycles in Newbury and Reading produced 5.1 and 9.6 per cent rates respectively.

The tendency of committees to block rather than facilitate applications indicates their lower tolerance of development. The 'low-growth' attitudes prevalent in Central Berkshire are thus reflected in the blocking of residential planning application. By blocking an application a committee accepts, or attempts to ignore, the risk of an appeal to the DOE against a decision which their own officers consider dubious or even untenable. The consequence of consent granted on appeal is often a development with fewer conditions attached than had been originally proposed by the officers.

In both the Bracknell study and the three-district survey, a higher level of blocking for new dwelling applications was evident. Clearly, residential development has been one of the more emotive and political issues with which local representatives have had to contend.

There are two strategies used by planning officers with blocking moves. One course is to stress planning policies, principles and conventions, in short emphasizing the good planning basis for the recommendation. When carried out with conviction this is often successful and will prevent many challenges from weaker members, whose 'amateur' status (and therefore lack of planning knowledge) has been subtly reinforced by the 'professional' counter-attack.

The second approach, complementary rather than contrasting, is the threat of appeal. Where committees do reverse a recommendation of consent based on clear planning grounds, there is a higher probability of a successful appeal, since inspectors give very considerable weight to the planning merits of a case. In addition, the applicant is likely to stress the divergence of views between the officers and members, which increases the chances of appeal success. There are two related factors however which add persuasive force to the appeal threat. First, appeal proceedings cost local authorities time and money; particularly if the appeal is heard at a public inquiry. Second, there are also sound planning arguments to discourage the committee. Traditionally, when inspectors allow appeals they impose fewer conditions on the permission than would the planning authority had they granted rather than refused consent. Thus the successful appellant has a freer hand in developing the site and the planning authority has forfeited an opportunity to influence a development. This can be a powerful argument against blocking, particularly when the officers' view is supported by the chairperson and/or other policy upholder councillors.

The appeal threat undoubtedly has a restraining influence upon committees. It often makes a blocking attempt a straight choice between principles and pragmatics. Some councillors will vote for a refusal despite near-certain appeal defeat. They maintain that district councillors should take decisions which represent the view of the local council and not of Whitehall. As one interviewee commented, 'I consider it to be blackmail. The planning legislation is inadequate when it allows unanimous committee decisions to be overturned by a central government department.' The survey of councillors revealed an almost even split between those who never or very rarely approved applications simply because of the fear of a successful appeal, and those who did so quite frequently.

FACILITATION

In an area where anti-development attitudes predominate, it is rare for planning committees to approve an application for which their officers have recommended a refusal. The Bracknell survey indicated that facilitation was as common as blocking for non new-dwelling applications, but was very infrequent (eight cases in as many years) for new dwelling applications (Fleming and Short, 1984). Members were seldom minded to side with residential developers against their own officers. Three main factors can be suggested in facilitation.

The human factor The notion which is espoused here is that the needs of an individual can override even the most rigid of planning controls. The applications which are facilitated are invariably small and the moral argument is that the development should be allowed to proceed for the benefit of a particular individual or group. The recipients' needs are stressed, as usually they are invalided or elderly. Many councillors expressed sympathy with the situation of particular applicants, but the human factor was most readily translated into decisions by non-policy administrator Conservatives and in particular those who could be classed as people's agents.

The amelioration factor On occasion a committee will identify an opportunity to improve a local environment by overriding policy and the planning officer's recommendation. Usually the benefits which accrue are the removal of eyesores and the 'costs' are that a new maintained dwelling is erected. The officers have weighed up the merits of the case but rejected the application on the basis that policy should not be undermined, or that more benefits could be obtained if this particular application was rejected but the submission of another encouraged. As with blocking it is policy upholders who are most likely to defend the existing policy or stress the need to act consistently, and argue against facilitation attempts.

The parish factor In Central Berkshire it is rare to find parish councils supporting applications which planning officers subsequently recommend for refusal. The parish councils have tended to be very anti-development in their attitudes. Nevertheless, on those occasions where they do support an application against the officers' view they can make an impact, usually through the ward councillor. Here the ward councillor acts virtually as the delegate of the parish council.

Reasons for parish support for facilitating moves are unclear but effective lobbying by the applicant is often suspected. Local applicants, known personally to the parish councillors, are the most likely to lobby at this level. However this channel of influence upon district planning committees has also been exploited by a very small group of planning consultants and agents. They have recognized that district councillors usually defend their areas because of local opposition to development, and local support for an application therefore undermines their instinctive response of refusal. Lobbying of key parish members, leading in turn to pressure on the ward member, can be used as a means of thwarting the refusal intentions of district planning committees.

From limited evidence about planning committees elsewhere, Central Berkshire authorities may facilitate less than others. One case cited in the literature is Hove BC in Sussex, spanning a twelve-month period in 1981/2 (Ray, 1983a). Of thirty-five committee reversals of the recommendation, thirty were facilitating moves compared with only five blocking actions (Ray, 1983b). Facilitation (and its overtones of lobbying) may have been more common in Central Berkshire before, and shortly after, local government re-organization than in recent years. A post-re-organization decline in facilitation was evident in Bracknell (Fleming and Short, 1984). The introduction of the public committee, and the strength of anti-growth feeling in Central Berkshire since the mid-1970s, may have dampened the enthusiasm of councillors to publicly respond to the wishes of an applicant.

Modifying Rather than accept the officer's recommendation in its entirety or reject it outright, a councillor can attempt to modify it. These adjustments usually take the form of adding or deleting certain conditions, but also include the addition or deletion of reasons for refusal, and in the case of county or government matters, amendments to the observations.

As Table 4.8 (p. 152) shows, the Bracknell councillors were particularly prone to modifying actions, both actual and attempted. The short agendas for development control and the leisurely pace of the meetings encouraged committee discussion to become involved with the details of schemes. Committee tinkering and amendment was less frequent in Newbury and Reading, with about 10 per cent of all applications affected in this way.

Stalling A stalling decision can take one of two different forms. First, an item can be deferred to obtain more information about the application before making a decision, or to allow further time for negotiations between the officers and applicant. Second, the purpose of a deferral can be to hold a committee site meeting to gain a better understanding of the actual site and its surroundings.

Deferral moves of both types are more than merely administrative actions. Four motives for deferrals can be suggested.

First, the committee consider that they cannot make a decision on the application on the basis of the information before them. The deferral is intended to obtain the necessary information, to obtain adequate plans and drawings in some cases, or to permit a site visit. These deferrals, although apparently administrative, none the less reflect division between members and officers, who are prepared to grant or refuse the applications as they stand.

Second, and related, the committee hopes to achieve better terms before granting permission. The applicant has satisfied the officers but the committee wants further concessions. Occasionally, committee-prompted 'further discussions' can result in an acceptable development from an initial refusal recommendation.

Third, the deferral can be used for political purposes. Some members call for site meetings in response to pressure from their wards, especially from parish councils who urge the local member to 'bring the committee out here and let them see for themselves'. Non-committee members are more likely to adopt this procedure, and planning councillors often resent such requests, particularly if the ward councillor has not personally visited the site in question. From the local member's viewpoint, however, this procedure is a useful method of satisfying local concern. Individual objectors, community groups or the parish council can be assured that, at the behest of the local member, the district council has noted their concern and will inspect the site.

Fourth, the deferral is instigated with the intention of challenging the recommendation.

These cases often hinge on the interpretation of the refusal reason proposed by the councillor, such as 'detrimental to the local environment', or 'inappropriate use for a rural area'. The local councillor will suggest to the other committee members that the impact of the application can only be assessed on site, in the knowledge that councillors are generally more protectionist than their officers.

Applications deferred for site meetings were much more likely to be blocked or facilitated than were items considered at committee. This was despite the fact that a minority of site visits were motivated more by constituency politics than by serious intentions to challenge a recommendation.

It appears that the site visit is a partially successful strategy in achieving a reversal of officer recommendations. By taking the meeting out of the debating chamber members realize they have a better chance of swinging the committee away from the officers.

Discussion

Committee decisions are accompanied in the majority of cases by comments or discussion. Public statements by councillors in the public forum of the planning committee were related to three general themes; the councillor's constituency, criticism of higher authority, and, mainly in Reading, party politics.

Constituency speeches are motivated by the need to publicly protect ward interests. As with justice, ward defence must not merely be done but must be seen, and even more importantly, reported to be done. A straightforward refusal of an application is therefore sometimes insufficient. To maintain local support and to provide useful publicity a short speech denouncing the proposal is required.

This form of 'playing to the gallery' was admitted by a couple of interviewees but defended as an accepted feature of a councillor's role. Considerable emphasis is placed on the public accountability function of the planning committee; the public is entitled to see and hear its interests being defended.

The constituency speech is of greater importance where the local member is speaking against a decision which seems likely to receive committee approval. Local members need to be seen fighting strongly for local residents and distancing themselves from any decision which would be damaging to their interests. Place agent councillors in particular will put up a strong case for their area. An excellent example occurred in Wokingham DC, over the relocation of a proposed local centre within a major residential development. The local member made a vigorous speech attacking the proposal, which received enthusiastic applause from residents who had come to hear the committee discussion. Despite supporting speeches and votes from councillors of adjacent wards the refusal motion was comfortably defeated, but the local member had publicly done all that was possible to prevent a consent decision.

The second type of public statement is *aimed at higher-authority*: the county council or the Department of the Environment. Both engender district council frustration by granting planning consents despite the opposition of the lower-tier authority, and both have received heavy criticism over the Heseltown issue.

County council items sent to the district for their observations received considerable committee attention. The district planning committees are powerless to prevent such proposals by institutional means, and attempts to use the political channel to gain concessions have had only limited success. Public statements were therefore frequently made to express resentment and to distance themselves from the impending county decision. A typical comment made in these situations branded the county council, in particular its property department, as acting 'as badly, if not worse, than any private developer'.

Planning committee attacks against the DOE were focused on

its power to determine appeals against local authority decisions, and the consequences of the structure plan amendment. It was essential to convey to the public the difficult position they faced regarding possible appeals. A number of members often made statements in these situations, even if a unanimous decision was likely, in order to publicly explain their stance. In the context of ultimate central government control through the appeal mechanism, councillors adopted a mixture of attacks upon the DOE, public explanation of their reluctant decisions, and emphasis of the gains and conditions which would result from a pragmatic approach.

Political speeches at planning committee were generally restricted to Reading, with a few at Bracknell. Office developments provoked the most heated conflicts, although Labour councillors attacked the policies and attitudes of their opponents on a range of issues. Residential items rarely provoked such statements, since there was considerable all-party agreement over the principle of residential applications, even if not over the details.

Summary: little power but all the blame

The land use planning system is essentially a mechanism for balancing the demands of a host of competing interests. Local authorities operate the system, subject to central government oversight and direction, and councillors are the decision makers. Their role, as of planning itself, is an intrinsically political one: choosing between alternatives given particular circumstances and constraints. Unless widespread consensus prevails as to the best interests of society for a particular locality, the decision makers will be unpopular with those disappointed by planning outcomes. In Central Berkshire the dominant local views are in direct confrontation with market trends, property and construction interests and central government's policy of encouraging growth wherever it occurs. As one housebuilder commented, 'Councillors are in the firing line; they should expect to be shot at.' The responses of councillors to the situation in Central Berkshire can be considered at two levels: collectively as planning authorities; and individually as planning councillors.

The collective response

At the collective level, committees have employed a combination of four approaches: resistance to development; public deflection of blame; control over details; and planning gain. Outright resistance by district councils has been severely limited by the power of the DOE and even Berkshire County Council. Where there has been clear intent to override or circumvent district opposition, planning committees have invariably conceded battles over the principle of development (particularly residential). The county's challenge to the additional 8,000 houses was supported, and Wokingham DC formally resolved at one stage to refuse any allocation within its area until 1986, but councillors realized that these moves were little more than defiant gestures. Futile refusals of residential applications were generally avoided, unless a delay was considered clearly advantageous. Office development in Reading up until the 1983 elections provoked a slightly different response. The committee had little to lose by refusing outline applications, since no effective control would be lost, and persisted with the defence of over-provision of floorspace contrary to policy. Councillors, especially Labour and Liberals, hoped that appeal inspectors would eventually accept their case and thus provide more support for future refusals. Even when this did occur, however, briefly in spring 1983, inspectors sometimes considered that other factors outweighed the accepted over-supply situation.

The second response has been deflection of blame. Where applications are approved despite local opposition, and often the personal conviction of councillors, a committee must make clear its dilemma. Speeches are required condemning the impact of the application, stressing the reluctance of councillors and the force of the appeal threat. Explanation of the choice between principles and pragmatism is intended to demonstrate the responsible course of action adopted. This response is merely a public relations exercise to disguise district impotence, but is usually accompanied by one or both of the remaining approaches.

Planning committees attempt to exert influence over the details of many applications, whether by personal involvement with officers or applicant, at private meetings or at public committee. There is greater incentive to exert control, however, over applications which would be refused but for Marsham Street or Shire Hall policy. 'If a proposal cannot be refused it must be planned', so therefore councillors become involved with the layout, design, functional composition and, in the case of many of the 8,000 houses allocations, the precise location of development. The greater the influence of the planning authority, but particularly of councillors, upon an application, the greater the justification for permitting it despite local objections.

This approach is easily extended into the area of planning gain. All but one of the questionnaire interviewees supported the principle of extracting some of the costs imposed upon a community from the developer. The extent of planning gain varied widely, both between districts and between types of development. Provision of adequate open space, new footpaths, 'village greens' and offsite drainage or highway improvements were commonly achieved on residential sites, with retail units, community centres and accommodation for elderly people a feature of some planning gain agreements. Office developments, particularly in Reading, were frequently subject to demands for some return to the community in exchange for the (often reluctant) granting of planning permission. Commuted sums of money to compensate for inadequate parking provision, and the inclusion of flats with office developments, were almost standard demands by Reading Borough Council. The demand to build in Central Berkshire, whether in the commercial, industrial or residential sectors, gives planning committees a strong position from which to bargain with developers, even if not to prevent growth.

The individual response

At the individual level the approaches of councillors can be considered in the context of the planning roles outlined earlier. All councillors perform an accountability function, which for council filler deadweights is virtually their only role. Only strong pressure from their usually peaceful wards is likely to provoke action, and if officers or influential councillors are opposed to their view, then little impact will be achieved. Lightweights are much more active in support of their constituents' views, and are generally more effective in the watchdog role. Whilst their influence may be limited, they are important channels for the expression of local views. Especially with inexperienced planning councillors, their simplistic or even naive approach may accurately reflect public opinion, but provokes contempt from developers and even some officers. Policy upholders by contrast are more likely to stir up local feeling, particularly among parish councils and community groups, by their stress upon consistency and responsible decision making. Concern for local objections is often displayed, but the response will be directed more towards control over, rather than refusal of, development.

Place agents have a clear response to the growth situation in Central Berkshire: attempt to refuse as much as possible, while demanding control and gain from all applications within their area. Particular individuals can become very efficient place agents; one councillor in Newbury was responsible for over one-third of the blockings during the eighteen meetings. Finally, the policy directors have the greatest impact, both at policy and development control levels. Their power can result from personal influence and/or political circumstances, and through the position of committee chair, may also extend to considerable control over the officers. Their capacity for early involvement with both applications and policies can effect a subtle influence upon planning within an authority.

In the context of DOE-encouraged growth pressure, councillors are frequently unable to translate opposition to development into planning outcomes. They do block applications with some regularity, overturning recommendations from officers who are themselves more defensive than colleagues in less buoyant areas. Their impact upon major decisions is limited, however, since, rather ironically, there is greater scope for facilitation than blocking in the planning system. This can be seen in the more pro-development actions of Reading Conservatives.

The influence of councillors is principally directed towards modification, control over development and planning gain. They are forced to work within the system, and the justification to local communities of their position is dependent more upon the developments they permit than those they refuse.

Politicians represent political, policy and place interests. In the next two chapters we will discuss two agents which articulate place interests. Parish councils and residents' groups are major constituents of local planning.

5 Parish councils

'a vital element to democratic local government'
(Redcliffe-Maud)

There is another layer of government below the district councils in much of rural and suburban England, the parish or town council: we will use the term *parish* to cover both. In Central Berkshire there are forty-five parishes which vary in size from 210 hectares to 3,888 hectares and in population from eleven to 48,750. In this chapter we shall consider the role of parish councils generally but particular attention will be paid to their involvement in land use planning. The raw material of this chapter was a survey of forty-three parish clerks and a more detailed questionnaire survey of twelve parish councillors (see Appendix 4).

As a general introduction to this chapter it is instructive to note the level of development undergone by the parishes of Central Berkshire between 1974 and 1982. We can simplify the analysis by using the two criteria of population density and relative level of planning application to construct a typology of parishes (see Figure 5.1).

First, there are the 'developed' parishes. These are characterized by having high population densities and high rates of application submission. The developed parishes include all the town councils and some other parishes to the east and west of Reading. Although these parishes have high levels of existing development, they are still recipients of large numbers of applications for ten or more dwellings.

Second, there are the 'stable' parishes. They have low population densities and low numbers of submitted applications. Some (e.g. Remenham) are protected from development by Green Belt designation, and others to a much lesser extent by their status as Area of Outstanding Natural Beauty (AONB) (e.g. Ashampstead). But others are effectively protected by ownership constraints: the owners of one very large estate stretching over

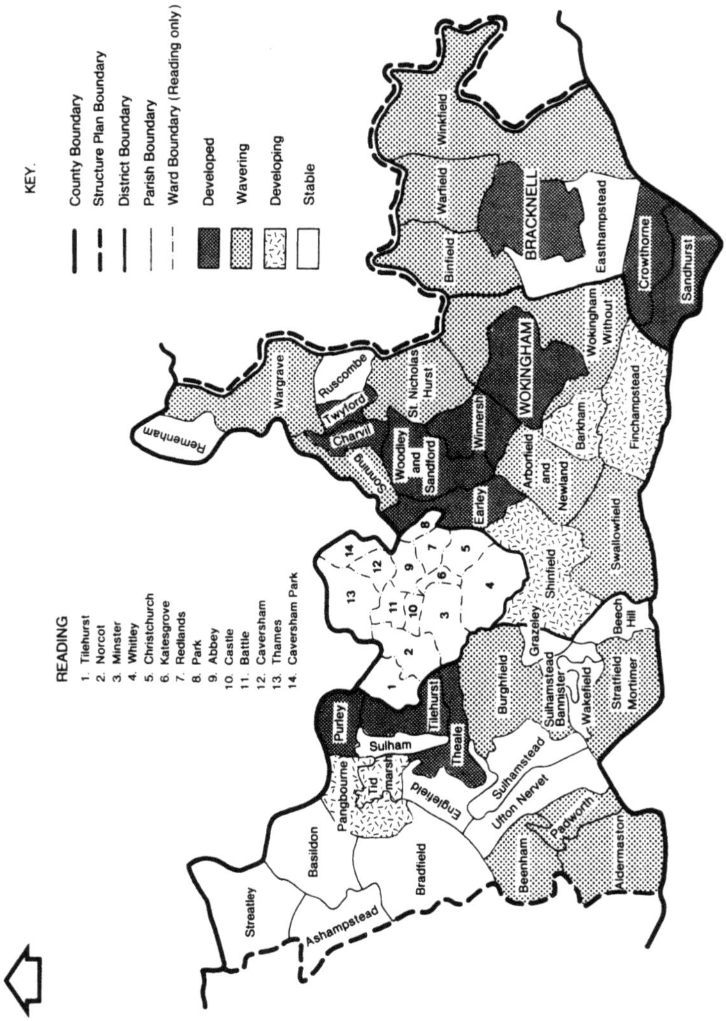

FIGURE 5.1 *Typology of parishes within Central Berkshire*

South East Newbury District have resisted many pressures to develop.

Third, and a little less distinct, are the 'developing' parishes. They are characterized as having medium population densities but high development pressures. These parishes form a string south east from Reading to Sandhurst and are also found to the west of Reading.

Finally, and most interestingly perhaps, are the 'wavering' parishes. They have low population densities but medium rates of application submission. In many respects, these parishes are those which are most caught up with the fight against development in Central Berkshire. They have moderate numbers of applications for ten dwellings or more but often a high proportion of these are refused. In many instances, developers are testing the ground in these parishes by obtaining fairly large sites and submitting applications for them despite regular refusal. Indeed, the status of some of these parishes may already have changed in the wake of the recent allocation of land for 8,000 houses in Central Berkshire. Burghfield and especially Warfield have been identified as major development sites and should the developments proceed, these parishes must enter the 'developing' category.

The above four-fold classification offers a useful baseline from which to view planning and the parishes. A cycle of development (from 'stable' through 'wavering' and 'developing' to 'developed') is implied but a longitudinal study would be necessary to verify if population densities and application pressures consistently follow the pattern suggested.

Parish organization

Origins

As the lowest tier of local government, the parish council is also the oldest. The parish unit derives from early church administration but the non-ecclesiastical, administrative functions of the parish emerged around the early seventeenth century and over a long period of time the parish gradually took on more functions.

The 1888 and 1894 Local Government Acts created a new structure consisting of the county, the borough and the (civil) parish. Parish councils could only be formed where a population threshold of 100 was reached, some powers (such as those over highways) were lost, but parish activities were entitled to be funded by a proportion of the rates. The powers of the parish were effectively limited to minor local issues such as footpaths and burial grounds.

The local government re-organization of the early 1970s reformed the whole system but it was recognized that parish councils could still perform a useful role. They were seen as 'a vital element to democratic local government . . . to focus opinion about everything that affects the well-being of each community, and to bring pressure to bear on the responsible authorities' (Redcliffe-Maud, 1969, p. 95).

The power and influence of the parish councils were enhanced by the 1972 Local Government Act. Under this Act the parish tier of government was entitled to have a council even if the parish population was less than 150. In any event, every parish was to have at least one annual meeting to discuss parish affairs and the councils were to comprise of members (the number dependent upon the population size) who were elected every four years. The functions and responsibilities of the parish council are shown in Table 5.1.

TABLE 5.1: Responsibilities of parish councils under 1972 Act

Communications
This includes the opportunity for the local council to provide car and cycle parks; lighting rights of way, footpaths and bridleways; seats and shelters; postal and telephone facilities.

Health
Includes public conveniences; water and drainage; sewerage; warehouses and launderettes; litter.

Outdoors
Relates to provision of allotments, sports and recreation facilities; roadside verges; open spaces and greens; fields.

Various buildings
Includes halls; indoor recreation; public clocks.

Entertainment, arts, tourism
Relates to provision of above or accommodation for activity.

Notifications
Planning applications; byelaws; cemeteries; particulars of land ownership.

Powers relating to the dead
Burial and cremation; mortuaries; churchyards; war memorials.

Education
The parish council can be the 'minor authority' relating to a county primary school.

General
Such things as charities; cooperation with voluntary bodies; and so on.

Committee structure

Parish government is the one level of local government which has a realistic choice as to whether it attempts to carry out all its business through its full council or whether it delegates responsibility and workload to a number of specific committees.

In Central Berkshire, thirteen (30 per cent) of the forty-three councils providing information had deemed it unnecessary to use a committee system. As demonstrated in Figure 5.2, most of these parishes were to be found in the rural west of Newbury East District. In part, their disuse of a committee structure stemmed from the small numbers of councillors in these parishes: each of the thirteen 'uncommitteed' councils had fewer than ten councillors. With so few councillors on the full council, the creation of even smaller committees was often thought inappropriate. However, there were another six councils with fewer than ten members which did utilize committee structures. But all of these six councils were in more (sub)urbanized parishes. Therefore a strong correlation appears to exist between use of committees and (a) number of councillors and (b) rurality. There are of course other highly important factors: the concentration of uncommitteed parishes in Newbury District might suggest a strong neighbourhood effect, while one parish clerk quite explicitly tried to prevent the formation of committees following the advice of his predecessor – 'don't let them start committees, they'll go on forever and you'll get nothing done.'

Parish council priorities can be partially indicated by the concerns of the committees which has been created. The most frequently occurring committees were those which were concerned with parish council property. Many councils owned recreation grounds, halls or allotments. To administer them, committees were formed in many parishes. General finance committees were very common in town councils but were seldom utilized elsewhere. Other issues which also merited concern included footpaths (mostly in rural areas), street lighting, general amenities and, in the wake of Dutch elm disease, trees. One of the more curious committees was an 'Emergency Planning Committee' set up to investigate what provision could be made for flooding, nuclear war and other traumatic events.

As shown in Figure 5.2, the plans, planning, or development committee was very common. However, it might be misleading to equate its ubiquity with its perceived status. Most parish councils have virtually a continuous inflow of planning (development control) matters on which they are invited to comment. To make comments and to do so efficiently is most easily done through

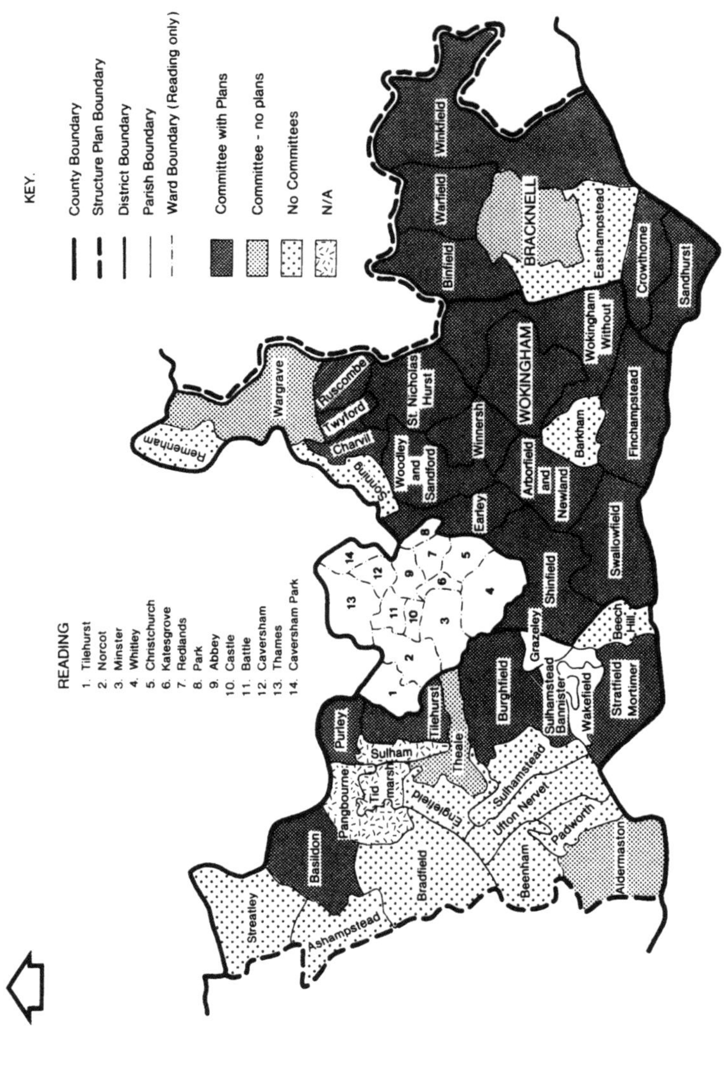

FIGURE 5.2 *Committee structure of parish councils in Central Berkshire*

committee. The practice of passing the applications from councillor to councillor appears to be dying because of the difficulty with which a council view is formulated; at least one local chief planning officer has taken a strong stand complaining about uncoordinated comments gathered in such a fashion.

The perceived status of planning is therefore not easily assessed. Certainly, it is seen as important but perhaps not to the extent to which is reflected by this pattern. Of the twelve councils which our interviewees represented, one had no committees and two others had no planning committee. When those with planning committees (nine) were asked to rank all the committees of the council in order of importance, four ranked the planning committee first or joint first. It was evident that the town councils did not rate the importance of their planning committees very highly – they were much more interested in the committees which controlled the purse strings. Having undergone considerable development, planning was something of a dead issue. But in the parishes north of Bracknell, the planning committees were rated very highly. Overall, however, there was an acute awareness that parish councils did not decide planning matters, they could merely comment upon them.

The clerk: the person with the reins

Parish councils are no different from all other levels of government in the UK in that the elected members are served by officers. However, the affairs of parish and town councils are so restricted that generally the part-time employment of a single officer (the clerk) is sufficient to service the needs of any parish council. In fact, some clerks service several parish councils and in Central Berkshire one clerk served six different councils. Given that many parish councillors may be naive of procedures, the clerk occupies a powerful position, a position which is further strengthened because (s)he acts alone and is not directly responsible to other officers.

Clerks sometime live in the parishes in which they serve (in this sample four of the ten) and councillors often consider this useful to ensure clerks' commitment to the parish. Since clerks generally have other jobs, the status of their posts elsewhere can be a useful attribute – many hold positions in district and county councils and can thereby be influential in giving their parish council an extra hearing at the higher government tier and in providing fast, reliable and often detailed information about relevant activity in the higher authority.

In inquiring about the relationship between the chair and the clerk, only one chairperson would admit to the existence of any tensions. Nearly all chairpersons (eight of the ten) claimed that their relationships with the clerk were of the highest order ('very good'). Clear feelings of deference were evident on the part of many chairpersons who were grateful to have a voice of authority to regulate the activities and discussions of councillors. With their experience and detailed knowledge of procedures, many clerks chipped in to redirect or stifle certain discussions. In issues such as planning which are quite technical and bound by superimposed codes of conduct, the restraining influence of the clerk is probably considerable.

The 'centrist' influence of civil servants has long been recognized at the highest levels of government. Such influence is also readily apparent in parish government where clerks frequently seek to minimize their workloads thus discouraging some new initiatives and restraining councillors from going beyond conventional boundaries.

Parish flamboyance – the parish precept

A significant force in parish politics is that which attempts to keep the parish rate low. One of our sample categorically stated that his prime motive in entering local politics (he was also a district councillor) was to keep the rates low. The reluctance of councils to raise their precept is one of the factors which devalues the potential of parish government. Their reluctance to collect money inevitably restricts their influence and helps to keep the profile of parish government low.

In 1983, the rating imposed by Berkshire County Council was 142.8 pence in the pound. The rates in the districts excluding Reading varied from 9.7 pence in Bracknell District to 11.1 pence in Newbury district. Beside these the rate requirement of most parish councils was paltry. In 1983 in Central Berkshire, one parish (Easthampstead) did not even have a clerk and collected nothing, but most parishes collected up to 3 pence in the pound. The town councils imposed the heaviest rates and the-then-Labour-controlled Bracknell Town Council imposed by far the largest rate: 7.2 pence in the pound.

The total amount collected from such rating of course varies considerably from parish to parish. A few examples illustrate this: Bradfield, a rural (stable) parish, having a precept of 3.1 pence, collected almost £2,500 per year; Tilehurst, a suburban (developed) parish, with a precept of 3.5 pence, collected approximately £58,000; and Bracknell TC, when its rate was 6.6 pence, collected over £600,000.

The disposal of these sums of money takes various forms. Money is directed towards the statutory duties which parish councils undertake (e.g. street lighting and footpaths), and the maintenance (or indeed purchase of) parish halls or recreations grounds often takes a large share. Parishes may also collect what is termed the 'free two pence' which may be used by the parish for matters which are in the interest of all or some of the inhabitants of the area. In the context of our study, the direction of funds from the product of the free two pence towards the costs of publication of a planning document is of particular interest. The document *Goodbye Rural Berkshire* and its findings will be discussed in detail in a later section.

Parish politics

Electoral enthusiasm

'I only stood to make up the numbers. If I'd known 'x' was standing, I would have withdrawn.'

(A rural parish council candidate)

The lack of interest in parish government often causes concern. There are two main ways in which parish council electoral interest may be monitored: first, by percentage turnout of the electorate, and second, by the number of candidates seeking office.

Although turnout figures are not available in any comprehensive fashion, some occasional and anecdotal evidence does give an indication of electoral activity. One electoral registration officer in Central Berkshire considered that turnout at parish council elections averaged between 20 and 25 per cent of the voting population. But the range is very wide: on some occasions the figure is very low (a 7.5 per cent turnout was recorded at Twyford in 1982), but it may be higher (35 per cent to 40 per cent) when parish and district council elections are held on the same day.

Parish council elections are due every four years. In Newbury DC and Bracknell DC the district and parish council elections occur simultaneously every four years, but in Wokingham DC one-third of the district wards and parishes hold elections on a staggered, cyclical basis three years out of every four. However, it is not always necessary for the electorate to cast their vote. Elections were necessary for only twenty-three (53 per cent) of the forty-three parishes which had councils. Four years previously the figure had been twenty-one, and so there had been a marginal increase in election activity. Three major clusters of uncontested council elections were evident (Newbury District, SE Wokingham District and NE of Woodley) and there may well have been some form of neighbourhood effect in parish council disinterest. However, when viewed in relation to the development pressures, it was clear that electoral activity was least in the rural areas and greatest in urban areas (Table 5.2). This is not to imply that development pressure was a major direct influence about parish activity, but rather that the small conservative rural parishes tend to have a low level of politicization, dislike the idea of explicit conflict, and generally find nothing unusual with the quotation at the beginning of this section. In contrast, the more urban parishes and particularly the town councils, with their few extra powers, attract a higher number of candidates and maintain a greater level of interest. It is also worth noting that in the more urban parishes there is a greater tendency to use the parish council as a stepping stone to the district council.

TABLE 5.2: Contesting of parish council elections in Central Berkshire 1979–83

	Contested (at least in part)	*Uncontested*
Stable	3 (25%)	9 (75%)
Wavering	9 (64%)	5 (36%)
Developing	3 (60%)	2 (40%)
Developed	8 (67%)	4 (33%)
Total	23 (53%)	20 (47%)

Throughout Central Berkshire, then, the level of enthusiasm at parish council elections is not altogether inspiring. Nearly half did not require elections, and just over a quarter (eleven parishes) actually showed a shortfall of candidates. The Redcliffe-Maud Report of 1969, which revealed a national shortfall of about 20 per cent, remarked that such a situation was unique to parish government and was due to the absence of organized party politics. We shall now consider party politics at parish level – both in terms of the general survey and in terms of the smaller interview survey.

The party line

'There used to be *no* politics in the council – we were all Conservatives.' (Independent/Conservative parish councillors)

The stance of the Independent has a long tradition in British local politics. Before (and even shortly after) local government re-organization, many district councils in response to queries about their party structures reported that they were 'non-party' or 'non-political'. However, with increasing political sophistication, the rise of the community-politicking Liberals, and the fusion of urban (Labour) and rural (Conservative) areas at the time of local government re-organization, the viability of the Independent withered. The Conservative party had to organize itself more explicitly and any Tory-sympathizing Independent might have to stand against a true Conservative candidate or might be encouraged to stand under the party banner by being offered a senior committee post. By and large the Independents were and are closet Conservatives – the Independent/Liberal does exist in relatively small numbers but the Independent/Socialist is exceedingly rare.

Today, very few Independents exist on the district and county councils of Central Berkshire but they are to be found in force on the parish councils. It is difficult to identify where party politics play a part in parish councils (the extremes are easily classified but there are many areas of doubt). Figure 5.3 gives a broad indication of the status of party politics in the parishes of Central Berkshire as gleaned from the responses of parish clerks.

As Table 5.3 demonstrates, the link between party politics and contested elections is marked. It also emphasizes the prevalence of the Independent – an element which was borne out by the interviews. Even where party labels were evident at the parish level most of the parish clerks contacted played down the role of party:

TABLE 5.3: Party politics and election contests in parishes of Central Berkshire

	Party	*Non-party*	*No reference to party*	*Uncertain*
Contested	8	10	4	1
Uncontested	0	13	5	2

> 'party politics do not enter into our parish council matters – much to my delight' (parish council clerk)
>
> 'our members have no party affiliations (at least not as parish councillors) and long may it remain so'. (parish council clerk)

It was often expressed that parish council activities were too trivial to invoke party differences. The absence of party, it was thought, reduces unnecessary and unproductive conflict. However, the absence of party also causes a reduction in

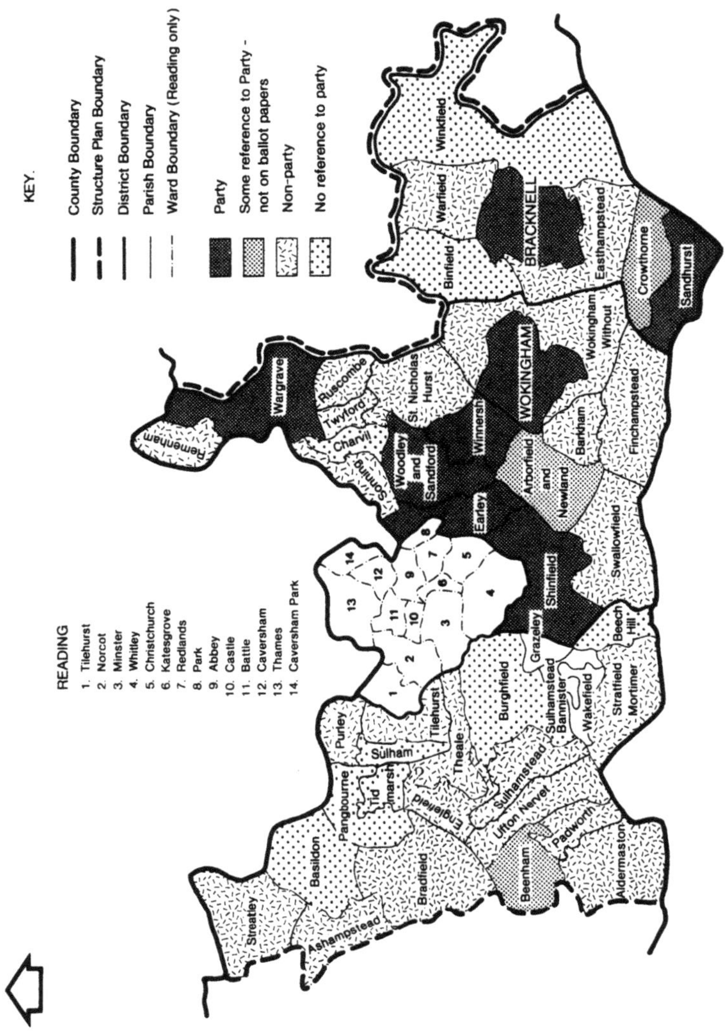

FIGURE 5.3 *Parish politics*

considered argument and often perpetuates the status quo by militating against radical change of any kind. There is no doubt that many chairpersons act to stifle any division along party lines:

> 'I've made it quite clear party politics are out.' (chairperson of a parish council)

> 'Some years ago, we had a bloody Labour chap who put himself and eight friends up as a Labour group. So then we put a pamphlet around the village saying we are *non*-party political. He got in – he was a damn nuisance but he did learn. He tired of the parish council and left the village.' (planning chairperson of a rural parish council with both 'Conservative' and 'Liberal' Independents)

Undoubtedly, the absence of party politics from parish government has excluded many elements of the community and has led to a domination by traditional Tory personages and those who like to become involved in small ways in village life. The 1983 election results for the parish councils of Bracknell DC readily demonstrates the composition of councils (Table 5.4). Even where non-Conservative or non-Independent candidates did gain seats there often appears to have been some tactical voting amongst Conservative voters to try to ensure the election of least worst opposition councillors:

TABLE 5.4: Candidates and elected councillors in the parishes of Bracknell DC, May 1983

	Total candidates	*Elected candidates*
Independent	54	45
Conservative	43	43
Liberal/SDP Alliance	33	6
Labour	41	5

> 'Some of the villagers said they would never vote for any Labour candidate. But I said they should vote for two of them to keep the two main Labour activists out.' (an Independent parish councillor)

Representing the parish: the parish councillors of Central Berkshire

Public representation at any level involves only a section of the population. Whereas in the higher echelons of government this has caused little anguish, at the level of local government it has caused much concern (Maud report, 1967). At the level of parish government it is generally thought to be particularly acute, because even though these councils have very few powers, they should in theory be closest to their electorate. As has been demonstrated, party politicking is relatively slight and has kept the profile of parish government low. The popular public image of parish councils is not high and they are often thought to be dominated by the affluent and the established 'old guard'.

Since our study involved only the interviewing of chairpersons (of the council or of the planning committee), definitive statements about the characteristics of all parish councillors cannot be made. None the less, the characteristics of these chairpersons are interesting and illuminating in their own right.

The parish councillors whom we interviewed were middle-aged or elderly. From casual knowledge of the composition of various councils in the area, there appear to be very few councillors in their thirties and fewer still in their twenties. The perceived and actual age structures in parish government are mutually reinforcing. Any young political activist is unlikely to be inspired by the vibrancy of the local parish council ('we deal in such tedious, trivial, domestic matters' reported one councillor with no particular *angst*). Although it is used by some as a route (intended or accidental) to the district and county councils, it is by no means a necessary route since the party involvement at the parish level is usually implicit or absent rather than overt.

Despite the lack of electoral enthusiasm, a degree of local credibility, usually obtained after a long period of residence in the parish and probably some activity within voluntary organizations, was normally a prerequisite for prospective parish councillors. On moving to an area, few councillors presume to obtain a seat on their parish council immediately. In fact, time lags of between ten and twenty years of residence and candidature were quite common amongst the sample.

Although our sample was predominantly male, the sex distribution of rank-and-file parish councillors was much more equitable. However, the fear was expressed by one woman that parish councils were coming to be dominated by the 'articulate woman'. Although she classed herself as one such person, she was not comforted by the trend and was concerned that parish councils in the area were coming under the control of wives of the professional immigrant classes. It was not the demise of the authority of the previously well-established male-dominated, agriculturally based community that was regretted but the fact that the council's former unrepresentativeness was merely being replaced by another dominant group.

The occupations represented by our sample certainly indicated the strength of the professional and managerial classes (see Table 5.5). Posts of authority are bound to attract professional administrators, but other anecdotal information suggested that the occupation types shown in Table 5.5 were a fair representation of the total council. Typical occupations tended to be that of insurance brokers, accountants and managers in the service industries.

TABLE 5.5: Occupations of parish councillors

Retired	2
Director	1
Managerial	3
Professional	5
Skilled	1

In the parallel study of district councillors discussed in the previous chapter, a typology was formulated which cross-tabulated motivations against roles. The re-adoption of that typology seems appropriate here even though parish councillors are excluded from many of the original categories and tend to cluster in a particular segment of the cross-tabulation. The absolute figures are shown in Table 5.6.

TABLE 5.6: Roles and motivations of parish councillors

Motivations		*Roles*			
		Policy administrator	Place agent	Council fillers	Total
Hard gain	Vested interest	–	–	–	0
Soft gain	Para-careerists	–	–	–	0
	Status-ticians	–	1	–	1
	Social councillors	–	–	–	0
Professional politicians		–	–	–	0
Paternalist/ maternalist		1	2	–	3
Activists		1	2	–	3
Get-involved		–	5	–	5

The motivations which were identified were as follows:

(1) Those seeking *hard gain*. Such councillors had vested interest in particular outcomes and might further their self-interest

through corrupt and underhand methods or entirely through legitimate channels. While some parish councillors may have sought office to protect their village and thereby their property values, the presence of those who are out to seek rather than to maintain material gain is probably slight. If such is their motivation, they will abruptly discover that the power of the parish council lies principally in its role as a pressure group rather than as a decision maker and is therefore not an efficient medium to obtain material benefit.

(2) Those seeking *soft gain* in the form of a social outlet ('social councillors'), self-status enhancement ('status-ticians'), or career substitution ('para-careerists').

As a social outlet, the post of parish councillor offers more tangible rewards. Unlike district councils which have their own offices and sometimes bar, parish councils have few facilities except perhaps the local village hall. Opportunities of using the parish council meeting place for recreational purposes are slight but the post of parish councillor does offer many opportunities and excuses for social encounters in and around the village. Undoubtedly, this is relished by many.

Similarly, the obtaining of a parish council seat may give some councillors a feeling of self-importance within their local community but that is unlikely often to be a primary motivation. There can be little doubt, however, that the small degree of status bestowed by the post is often enjoyed (though seldom admitted to) by councillors.

It has been noted elsewhere (Rees and Smith, 1964) that some councillors (often with unrewarding full-time jobs) consider that their position on the parish council offers an opportunity to develop a subsidiary career. However, the level of sophistication and the lack of (perceived) power of the parish council is not likely to encourage councillors seeking para-careers. Despite or perhaps because of the fact that many of the councillors interviewed already held managerial posts, there was a hint that they enjoyed using their expertise within the local community. Rather than substituting for an unrewarding career, a post as councillor allowed for further development of personal abilities. Such a motivation, however, was more likely to be of a subsidiary rather than a primary nature.

Aside from personal considerations, there are other more externally orientated motivations. In the typology of district councillors, we identified three main motivations which appeared to have some connection with party affiliation, philosophy and culture.

The *paternalists* or *materialists* were very frequently found amongst the rural Tories (even though the party label was often not employed at the parish level). As the local gentry or pseudo-gentry, they considered it their duty to become involved in the affairs of parish and village. Their role is seen as maintaining the status quo but so internalized is their view of society that they consider the status quo as being the 'natural order' of things. This group of councillors usually espouse traditional Conservative values and seek to protect their physical and social environments. They usually take a 'no-growth' stance in planning matters and are seen to support (or offer no resistance to) only the smallest of proposed developments.

The *agitators* and *activists* are motivated to become parish councillors by their desire to change existing regimes. They usually arise in response to specific issues and in our sample we found two who had responded to the threat of greater residential development.

Get-involved councillors form by far the greatest proportion of the councillors we interviewed. The archetypes are people who value and often idealize village life and become involved in many of its organizations. Their growth of involvement in village life in general is gradual – from membership of local residents' groups, they may be asked to help administer some voluntary charity and eventually be asked to become a candidate for the parish council. But to become a parish councillor it is generally considered better to be asked than to take the initiative and to forward oneself.

The roles of councillors may be cross-tabulated against the above motivations. In the survey of district and county councillors, we identified three which are again relevant.

First, there are those who formulate and try to administer policies – *policy administrators*. Since the powers of the parish council are perceived to be so weak and so few, there are few who can legitimately be categorized thus. Policy initiatives are minor and generally the parish council is merely reactive to events. But there are some notable exceptions. In planning terms, the joint publicity by Warfield and Binfield Parish Councils of the booklet *Goodbye Rural Berkshire* was an impressive initiative. One of these councils also helped establish, but remained independent of, a neighbourhood association. Although they were in receipt of outside help, these initiatives were undoubtedly largely attributable to policy administrators.

Most parish councillors are perfect examples of the *place agent*. The role of place agents is to safeguard their own areas from perceived threats. Since parishes are normally centred around villages, place agency is a frequent motivation and a role. Indeed, it is no coincidence that the word 'parochial' has come to imply a strong sense of narrow-mindedness and tunnel vision. At their

Block 15 The composition of two parish councils

These thumbnail sketches are based on anecdotal information supplied by a few parish councillors.

The narrow-based (stable) rural parish council consisted of eleven members, with a fairly equal male–female split. It contained no business people but three local teachers from administrative occupation families, two farmers who were 'sensible but not live wires', a local builder 'who takes no real interest', a lady 'with a gift for upsetting people' and a few other quiet and undistinguished people.

The broader-based village/suburban (wavering) parish was of quite a different nature with representatives from each of the quite distinct communities within the parish. Representatives came from a tenants' association, a residents' association, and a multinational company ('we treat them as a community'). Others represented less specific interests: a 'Tory party representative', a wife of a former mayor of the nearby town, and the rest were 'old-established villagers'.

worst, place agents are protective of their immediate environment and give no consideration to events even just outside their parish boundaries. Hence, in planning terms, the worst of the place agents have come to adopt an attitude of 'no development here, we're special and vulnerable' but give no thought to development elsewhere.

And finally, there are the *council fillers* – those whose position on the council has little impact since their contribution to debate and action is minimal. Their numbers, however, are lower than might be anticipated since parish councils are often at great pains to distribute the workload evenly.

The unrepresented parishioner

Certain types of parish resident are excluded from formal parish affairs. Few vote and as we have shown parish councillors themselves cover only a limited part of the social spectrum. The occupants of the small council estates of the rural parishes appeared to take no part in parish activities. In some ways, the parish council was perceived as a special club: 'The residents of the council estates of the 1950s and 1960s take no interest in the parish council. Instead, they run the football club and the Scouts and Guides' (a councillor of a wavering parish).

However, some parishes are more representative than others and a distinction can be drawn between the narrower rural-base parishes and the wide-based suburban parishes. Block 15 provides illustrative examples.

Public participation in parish affairs

Public attendance at parish council meetings is renowned as being notoriously low, but attendance is probably no lower than for district and county council public meetings. True, parish council meetings by being local are more accessible and perhaps their concerns more immediate to residents but it is a little curious that low attendances at parish meetings are frequently singled out while the upper echelons of local government escape such accusations of apathy.

While the monthly parish meetings seldom attract a public attendance which reaches double figures, the annual meetings are better attended. The monthly meetings are usually only attended by those with specific complaints or interest or involvement in particular items, perhaps (but not always) the local press, and in some parishes a hardcore of elderly attenders who like to keep in touch and to enjoy their monthly entertainment. Some parishes have tried to encourage, by informal and formal means, public participation at meetings but have been disappointed by the low response and the nature of the participation – usually the airing of grievances.

The annual meetings, though, are more vibrant; they may involve the setting of the *parish precept*. In the knowledge that a fuller attendance is likely at the annual meeting, consideration of major items is often withheld for that occasion.

Generally, therefore, the indicators which are available (electoral candidature, votes cast, and public attendance at meetings) strongly suggest a low level of interest in parish matters. But there are of course occasions when interest explodes and particular events or issues are the focus of considerable attention. The interest reflects not a concern with the notion of parish council administration *per se*, but rather demonstrates that the parish council can be a highly accessible arena (both in locational and political terms) for debate.

Parishes and planning

Since the 1974 local government re-organization, parish councils have held a prominent position in consultations over planning issues. Before 1974, there was consultation of parishes but it was neither comprehensive nor uniform. The 1974 reform gave parish councils the statutory right to be invited to comment upon planning issues. On making a request to their local planning authorities, parish councils were to receive notification of planning matters affecting their area. In Central Berkshire, the request

Block 16 A week in the life of a parish chairwoman

What does a parish councillor do? We asked one of our parish council chairpersons (of a stable parish) if she would indicate, in the form of a diary, when and how much time she spent on parish council duties. She did and it is set out below to demonstrate how one semi-retired woman coped:

'My parish council life seems to go along in a series of snippets. Snippets, somehow, do not conjure up a picture of real work.

Week July 4–9

Monday	
10 a.m.	Telephone call from elector worrying about repairs (carried out by district council workmen) to a wall. Are they in order? Go to see it and her. Find it's OK. (40 minutes)
Tuesday	
10–11.30	Gathering papers etc. for parish council in evening.
5 p.m.	Call from clerk about small point to do with meeting.
7.30–10	Parish council (first Tuesday of the month)
Wednesday	
11–11.45	Informal talk with a Charities Trustee who also signs a cheque.
12–12.45	Walking a bridle path to see why there are complaints that a horse can't get through.
8 p.m.	Clerk rings up to discuss meeting (20 minutes)
Thursday	
10–12	Working on drafts, correspondence, etc. Some telephoning.
7 p.m.	Available for judges coming to see Tree Planting Scheme near house.
Friday	
11 a.m.	Telephone rector about nominations to the Trustees of the Almshouses. They are appointed by parish council.
4.15	Telephone call from worried neighbour.
6.30–7.30	Site meeting at a difficult planning application.
Saturday	
10–5 p.m.	Conference at Reading University – Council for the Preservation of Rural England. (Saturday conferences are perhaps two or three a year. There are also about six or eight evening meetings on subjects of interest to PCs.)

I would not say this is a typical week. Others are quieter; nothing much happens but I do a good deal of reading on parish matters, or writing reports, bits for the parish magazine, letters, often in bed or very early in the morning. Then, again, a chairman's life is probably a bit different from a councillor's.'

seems to have been a formality and it appears that there are no parishes which are not currently notified.

The way in which parish councils are treated as consultees varies amongst districts and issues. In county policy matters, parishes are treated no differently from other participants in the consultation process, but in development control matters parishes do obtain information which is less accessible to other groups.

In consideration of development control terms, each parish council in the three districts of Central Berkshire receives a copy of every planning application (accompanied by the plans) for its locality. Each application and plan is one of a set of three or four which applicants must submit to the local planning authority – thus the planning authority merely acts as postman. Occasionally, however, when extra information is pertinent (such as applicants' letters) that information is also included – but this is undertaken on an *ad hoc* basis and at the discretion of the planning authority. In addition to the detailed information about applications in its own administrative area, most parishes also receive regular listings of applications for the whole district which are made available (usually at a cost) to any interested party.

While half (six) of those we contacted considered that the information on planning applications which they received was sufficient, four councillors considered it inadequate and another two had slight reservations and wished for specific pieces of information. Those who wanted more information were generally concerned with greater levels of detail – one councillor even claimed that the local planning authority was lax in not insisting upon the applicant providing all the information demanded by statute.

Interestingly, only one councillor volunteered that he wished to know the officers' views on each application. Because of the frequency with which officer recommendations are translated into decisions by planning committees, the lack of curiosity about officers' views was surprising. Doubtless, in particular circumstances, their views were ascertained by other contacts but in Newbury District and Wokingham District there was general ignorance of officers' views and this cannot have helped the parish councils to direct their energies to particular applications. In Bracknell District, however, procedures were different. By means of the fortnightly public listing, the officers' 'initial reaction' to a proposal was publicized. While the initial reaction might later be changed, it did give public and parish alike an indication of the likely decision on a proposal and thereby enabled objectors or supporters of applications to pay especial attention to initial reactions which were contrary to their own views.

Generally, however, those interviewed were content with the amount of information which they received from the local planning authority. None the less, that did not stop them from obtaining more information through their own efforts. Indeed, there have been several recent instances in Central Berkshire where parish councils have forwarded corrected information to the planning authority as a result of their own investigations.

The number of planning applications submitted for any parish varies enormously. Over the period 1974–81, the number of new dwelling applications ranged from 0 in Sulham Parish to 507 in Wokingham Town. Clearly, the constant throughput of planning applications can necessitate a considerable amount of work if a developed or developing parish wishes to exert its influence.

To formulate the parish council view, different parishes have adopted different modes of operation. One method is to establish the parish council view at the main monthly (or six-weekly) meeting. This can work well in small parishes with few applications and/or few councillors, but can become difficult in larger parishes. This method is also used in councils where planning is of low priority and where planning issues are barely discussed. Another method, though little used and often inefficient, is to circulate plans from councillor to councillor to make their comments which may or may not be forged into a single coherent view by the leader of the council. The circulatory method appears to be little used in Central Berkshire.

Of much greater frequency is the planning committee. The committee consists of between one-third and two-thirds of the total number of parish councillors. The allocation of councillors to committees is seldom systematic – parish councils being the organizations they are, allocations are usually made as the result of 'words in ears'. Anyone with any known expertise in planning finds little difficulty, and indeed can scarcely avoid, obtaining a place on the planning committee. Running in a phased cycle with the main monthly meetings, many of the Central Berkshire planning committees also met monthly. However, in Bracknell District, where a few parish councils took a very keen interest in planning, the committees met twice as frequently as the full council. And in Newbury District, where pressure had recently been applied by the district planning authorities to obtain comments more quickly from the parish councils, the committees met fortnightly. The planning committee therefore took up a large measure of council time in attempting to keep abreast of the input of planning applications.

There are two main channels through which parish councils may articulate their views: the professional channel and the political channel. Implicit in the Local Government Act of 1972 was the notion that parish councils would articulate their views officially to the local planning authority – in practice, this means the officers. Since most comments are committed in writing, the professional channel and the articulated comments will be considered together. But there is also the political channel – it is very influential, and in fact essential if a planning officer's recommendation is to be overturned. The message conveyed through the political channel may be identical to that conveyed through the professional channel but often it is subtly different and seldom penetrable. The political channel will therefore be treated separately.

The professional channel

Perhaps because of its relative simplicity and greater immediacy, development control items evoke a higher response from parish councils than do policy issues. This section will outline the articulated comments submitted to the local planning authorities on development control and policy items and then brief consideration will be given to direct parish council–district and county officer links.

The articulated concerns of parish councils in commenting upon residential development proposals

Most parish councils do take the opportunity to submit their views on development control matters to the local planning authority. In our interview survey, ten of the twelve parishes contacted claimed to comment upon *every* application. Many interviewees clearly believed that the credibility of their council would be increased if they took the opportunity to use fully their statutory right. Table 5.7 shows a range of comments from one suburban developed parish.

In commenting upon development control items, parish councils are not bound by any framework for reply. Their comments may be as long or as short as desired and need not be confined to simple statements of objection or approval. Indeed, without supporting information, statements of objection are likely to carry little weight. Because of their *penchant* for objection, however, simple statements of 'no objection' are likely to be interpreted as support for proposals.

Comments upon development proposals may usefully be classified as those which involve the principle (i.e. as to whether there should be any development on a site) and the detail (i.e. as to

TABLE 5.7: Parish council observations on planning applications

---

**Application no. 18779*: Erection of detached single-bedroom retirement bungalow – land adjoining 34 Culver Lane, and fronting Palmerstone Road, for Mr R. Hewitt – no objection, provided that affected neighbours are agreeable to proposals. It was noted that there was provision for car parking on the site of the application; these proposals render void the existing garaging arrangements for no. 34. There should be an associated requirement for arrangements for vehicular access to no. 34 so that there will be no regular on-street parking at this road junction.

**Application no. 18760*: Erection of site compound together with all necessary temporary structures and fencing – Plots 120–121, area 2 Lower Earley development for Taylor Woodrow Homes Ltd. – no objections except that there was concern regarding the proximity of the proposed compound to the open space area. There should be adequate safeguard to prevent an overflow situation whereby the open space may be used for storage purposes.

**Application no. 18804*: Construction of access collector road no. 7 and access way no. 8 – Swallows Meadow, Lower Earley for Bryant Homes Southern Ltd. – no objections.

**Application no. 18792*: Single-storey side extension to provide garage and new front porch – 5 Byron Road for Mr G. Chand – no objections.

**Application no 18701*: Erection of twenty-six flats – local centre site Area LC, Kilnsea Drive, Lower Earley for J. A. Pye (Oxford) Ltd. (Parish of Woodley) – the council are of the opinion that the application should be refused.

Redesignating this community use area for residential purposes will destroy the whole concept of a local centre. There should be no detraction from the original brief that community and commercial elements should be included at this location.

***Application no. 19007*: Erection of ninety-eight residential units with roads, footpaths and garages – area AA Lower Earley development for Bovis Homes Ltd. – the Council are of the opinion that the application should be refused.

There appears to be a lack of regard for pedestrian movement within the estate and exit from the estate. There should be a footpath system which will lead to Rushey Way so that pedestrians can cross to the district centre and also a link to Beeston Way to pick up the public transport route. Footpaths should be provided alongside the full length of the estate roads.

The open space area immediately adjacent to Rushey Way should have the benefit of suitable barriers to discourage children straying from the open space area on to the main traffic route at Rushey Way.

TABLE 5.7 – continued

***Application no. 18992*: Two-storey side extension to provide garage and utility room with bedroom and shower room above at 11 Launceston Close for Mr T. Disley – the Council are of the opinion that the proposals at the side elevation will affect the visual amenity currently enjoyed by the residents of no. 10 Launceston Close and there will be an intrusion into their privacy.

Sources: *Earley Town Council: Minutes of Planning Committee, 24 January 1983
**Earley Town Council: Minutes of Planning Committee, 21 March 1983

what the exact nature of a development might be) of a proposal. To try to identify the concerns of the parish councils, we asked them which issues their comments focused upon. While five blandly asserted that their comments were comprehensive, others were more discriminating and could see that they were pursuing particular issues.

Interestingly, two parish councils claimed not to be concerned with comments upon the principle of a development. One, a Green Belt parish, was quite confident that the local planning authority would enforce the restrictions upon Green Belt development and so the parish council could afford to direct its energies to the detail of likely developments – in particular this council was interested in the design of proposed developments. The other parish council was not in such a rigidly defined area, yet it felt that the local planning authority was predictable in its implementation of policy at least on the grounds of principle – this parish therefore concentrated on various aspects of the detail of a proposal.

In defence of their village, three parish councils (one was the Green Belt authority mentioned above) indicated the existence of a village envelope. The concept of a village envelope has no statutory force but is often used as an informal planning guidance. The envelope concept was introduced by the local authority planners to Central Berkshire in the early 1970s. It has been eagerly seized by many councils as a tool to prevent any substantial expansion of their settlements. A line drawn tightly around the built-up area of the existing settlement purports to restrict development – while infilling may be permitted within the line, there is substantial opposition to any development outside its limits. Even though the envelopes have no statutory force, they are quoted by the planning officers and, at appeal, DOE inspectors have taken note of them. One parish council, in fact, considered its major task to be the defence of its village envelope.

Density is often thought to be a contentious planning issue but, intriguingly, only one parish council volunteered that this was a major target of their comments.

Design and character of dwellings were major concerns of three parishes, one claiming in almost archetypal fashion that it fought 'out-and-out modernity'. It was interesting to note, however, that although individual applications are frequently fought on the basis that they were 'out of character' no councils which we encountered ever forwarded any positive policies on dwelling design and layout – their actions were mainly rearguard and defensive.

Finally, the general tenor of all our interviewees was anti-major development. Only a Labour-controlled town council explicitly stated that it wished to encourage residential development, and this was to be at the expense of office construction. But perhaps even more significant was the fact that only one parish (about to undergo major development) volunteered that it was concerned about ensuring adequate service provision for developments – the seeking of planning gain (in its most overt fashion) is seldom recognized as a facet of the planning system by parish councillors.

Parish council official comments on development proposals tend therefore to be negative. Their role tends to be of a 'blocking' rather than a 'facilitating' nature, i.e. where they disagree with the planning officers, they will tend to be objecting to rather than supporting an application. As will be seen in the section on policy, parish councils are generally reluctant to welcome any development for fear of being engulfed by more than they wished for. The effectiveness of their comments is difficult to judge. When asked to assess the importance which planning officers attached to their views, none claimed to wield a strong influence. However, it is important to recall the data in Table 4.4 (page 128) which showed that parish councils were rated third after officers and community groups by district councillors.

Parish council comments on matters of planning policy

Although parishes may become involved in the consideration of planning policy issues at the district level, the recent structure planning activity and the public participation exercise over the Heseltown issue has caused the major policy arguments to focus upon the county council, which has responsibility for strategic planning considerations. It is at this level that we shall focus attention.

From the Berkshire County Council's *Report of Public Participation* (1978) and from a report to the county's Environment Committee in December 1982 (Environment Committee, 14

December 1982, Agenda item no. 8, Appendix 2), it has been possible to construct a picture of parish council involvement (through written submissions) in the structure planning process and its subsequent amendment. In practice, the parish councils were treated like every other body in the public participation exercises in being invited to submit their comments on a number of documents and at meetings. The sequence of relevant publications and meetings associated with structure plan generation and modification were as follows.

(1) *The Issues Report* (June 1974). This 37-page document identified the context, scope, function and preparation of a structure plan for Central Berkshire.
(2) *The Report of Survey* (June 1976) gave much more detailed information after research had been undertaken in the wake of *The Issues Report*.
(3) *The Consultation Document* (May 1977) set out the policies and proposals being considered for inclusion in the structure plan.
(4) Public meetings followed the publication of *The Consultation Document* and were held in various centres in Central Berkshire in the early summer of 1977.
(5) *Land for 8,000 Houses: The Choices* (1982) was designed to give the public the chance to indicate where they would like the 8,000 houses located.

The pattern of parish council response to the various structure plan and 'Heseltown' consultations is given in Figure 5.4.

As many as seven of the forty-three parish councils made no response whatsoever at any stage of the consultation process. This absence of response was confined to the very rural ('stable') parishes. Evidently, they felt no particular threat to their immediate environment and so were disinclined to respond to the consultation invitations. They may well have considered that the best way to maintain the status quo in their parishes was to keep a low profile.

At the other extreme, only two parish councils made representations at every possible opportunity. Each of these parishes had already experienced a fairly high level of development and were not particularly threatened by future development; but it was interesting to note that one had as many as three dual representatives and the other had one very influential dual representative. It is probable that these dual representatives were instrumental in ensuring that their parish councils submitted their comments. Their comments were, in fact, quite detailed and indicated that they had a clear understanding of the planning issues at stake.

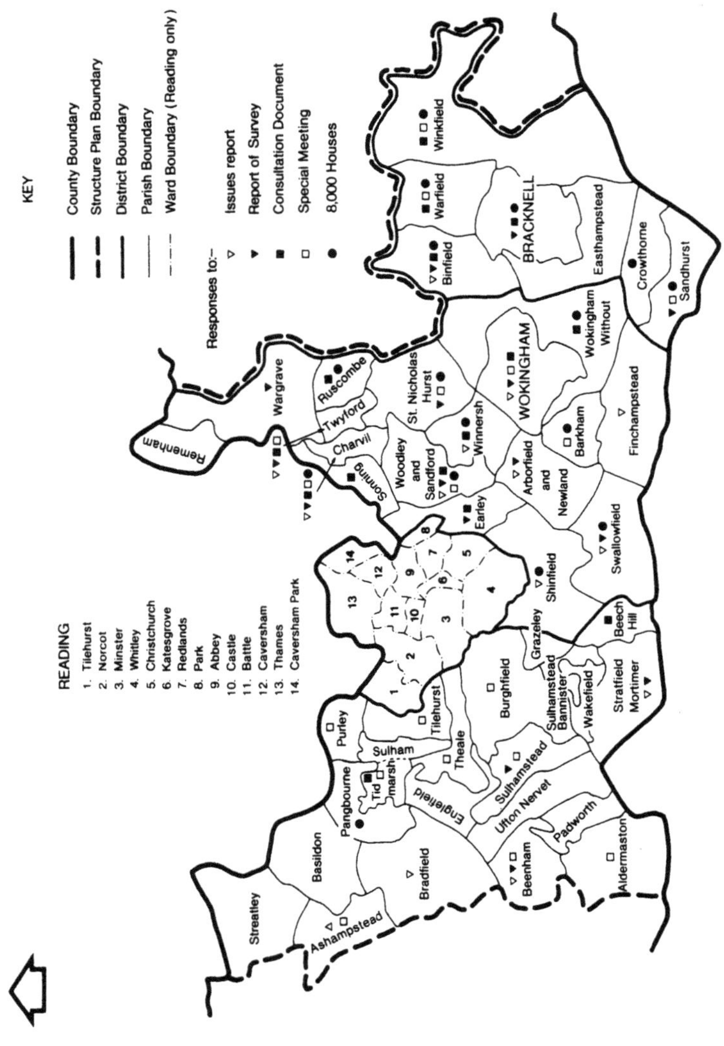

FIGURE 5.4 *Parishes and planning policy*

They reluctantly accepted limited growth but took the opportunity to request that local plans be made and that infrastructure and facilities (planning gain in its broadest sense) should be provided.

An interesting distinction can be made between these parish councils which responded only to the early documents (the 'early respondents') and those which responded only to the later documents/meetings (the 'late respondents').

The early respondents were galvanized into action at the first stages of the consultation process. They numbered five, four of whom were encountered in the survey. Three of the surveyed parishes recognized at a fairly early stage that their particular areas were not under threat from development. One situated in the Green Belt and another in the Area of Outstanding Natural Beauty considered that, although they were in Central Berkshire, the accommodation of growth in the whole area did not really impinge upon their parishes. The Green Belt parish saw that national and local planning guidelines were sufficient to protect the status quo of their parish and considered they could remain aloof from the general problems of the growth pangs of Central Berkshire. Though not so clearly protected by its status, the AONB parish did not feel particularly threatened, although it did recognize that development in a nearby parish might ultimately have a spillover effect. The third parish, in contrast, had no official protective designation but it had launched an early campaign to try to preserve its character. That campaign, it felt, had been successful and, although the parish council was still vigilant, it considered that there was no need to further the original campaign. The fourth early responding parish was of a different nature. It was a parish subject to development pressures and it did resent the development, yet it had made no recent representations to the county. An air of impotence and inevitability was evident (though it should also be noted that the councillor interviewed was rather naive of many planning procedures): 'I am strongly opposed to more development in the area but I don't think that whatever we do we've got enough strength. The district council has been compelled to come to heel by the county – we have to accept likewise. We haven't had much effect.'

Thus the early responding parishes were of three main types – those who had successfully negotiated the planning machinery, those who need not do so, and the disenchanted.

In very marked contrast were the late respondents. Aroused to action and response by the realization that the status quo of their parishes was under threat, these parishes had begun to take an interest in the planning process. Interestingly, but unexpectedly, only two parish councils had responded solely to the 'Heseltown' issue. While the 'Heseltown' issue did generate a new *public* awareness and reaction, very few parish councils were first motivated to respond to planning policy matters at this relatively late stage.

The Bracknell District parishes formed a large proportion of the late respondents. And, indeed, it is this area which has increasingly come under development pressure. However, there is little evidence to suggest that their early quiescence led to the subsequent development pressure.

The parishes' response to the Heseltown options

As a result of the DOE's modification of the Central Berkshire structure plan, the county planning authority was instructed to allocate land for a further 8,000 houses. A public consultation exercise was undertaken and, amongst the many other consultees, the parish councils of Central Berkshire were invited to respond to the Berkshire County Council (1982) document *Land for 8,000 houses: the Choices* (the Options Document). The Options Document identified five different combinations of possible development sites and consultees were asked to rank these options in order of preference or to recommend a different combination of development sites.

Written responses were received from seventeen councils but by the time many of them responded, the idea that the structure plan modification might be fought in principle had been widely aired. In the event, the parish councils used the consultation process to object to the structure plan modification in principle and to fend off development from their own parishes or nearby areas.

Of the seventeen parish council replies, no fewer than twelve took the opportunity to object to the allocation in principle. According to the Environment Committee minutes, two councils actually confined their comments to that issue alone and none explicitly supported the structure plan amendment. Only three parish councils appeared to accept that in future they might be prepared to accept development and only one parish (Crowthorne) agreed that it might receive its proposed allocation of 200 houses. For the rest of the parish councils, the consultation exercise was used to direct development elsewhere. Six councils identified a specific option – all of which excluded development in the parish concerned – and another five indicated more generally (i.e. without stating an option) where development might proceed.

It was interesting that no parish council appears to have used the consultation process to negotiate for a smaller slice of development. From interviews, it was apparent that many considered the best negotiating stance to be one of 'no homes here' – to admit to part of the allocation, it was thought, would show weakness and encourage the planning authorities to make a much larger allocation.

According to the Options Document, fourteen parishes were potential recipients of part of the 8,000 houses allocation. In the event, four of these made no official written response to the Options Document. However, an absence of response need not imply an implicit support of the allocation. Said one parish councillor whose council had not responded, 'What chance has a little parish council got against Whitehall and Shire Hall [the county council HQ]?' Although the four non-responding potential development recipients may have made their views known through less formal channels, it was intriguing to note that they all did in fact receive a proportion of the eventual allocation.

The impact of the parish council comments on the Heseltown options will never be known. Whatever the intent, the effect was to play one parish off against another and many councils resented this. To counteract this and to attempt to coordinate a united parish stance, the leading lights of two parish councils (which were threatened by development) instigated a collective response from the Berkshire Association of Local Councils (BALC). In line with the thinking of the two leading parish councillors (from Binfield and Warfield) the essence of that comment was to reject 'the whole concept of Central Berkshire as a major growth area' and to urge the county council to submit an alteration to the approved structure plan which would avoid the release of land for 8,000 houses.

From the structure plan consultations to the Heseltown consultation, a hardening of the general parish council anti-development stance is apparent. Whereas in the structure plan consultations the parishes may have been articulating their genuinely held opinions, the Heseltown consultations appear to have elicited a harder public stance of 'no growth'. Rather than negotiate for levels of development, the predominant attitude was to fight off all and any development.

Parish council links with local authority planning officers

It is usually the duty of the parish clerk to process the parish council's comments upon planning matters and to submit them to the appropriate authority. Apart from the regular exchange of planning documents, contacts with the planning officers of the local authorities are comparatively rare. Only two of the twelve parish councils contacted claimed to have other contacts with their district on a monthly basis. Contacts with the county are even less frequent and parish councils claimed never to have contacts with the county officers.

Links with district officers are facilitated by the regular district–parish liaison meetings. The frequency of these meetings vary: Newbury DC holds an annual meeting and, in addition, in 1983 convened a 'planning seminar' for interested parish councillors; Bracknell DC holds a bi-monthly meeting; and Wokingham DC convenes a district–parish liaison meeting twice yearly to discuss general issues. In all cases it is the district which sets the agenda and, by the admission of district officers and parish councillors alike, the meetings are little more than public relations exercises. Few of these meetings are dominated by planning issues.

Parish councils may on occasions request that a district planning officer attend one of their meetings. The district is not required to oblige and, while some requests are met, others are politely dismissed or discouraged. Even when they do attend, district officers make the point that they have come to listen rather than to inform, so clearly the usefulness of such meetings is limited. The districts are wary and generally attempt to distance themselves from the parishes. In defence of their self-distancing, the districts often claim that there is adequate opportunity for parishes to communicate their attitudes in writing and that, at the political level, either the district ward member is a parish councillor, or there is close liaison between the two, thus ensuring that the parish is adequately represented.

The district-organized site meeting is one of the major media through which parish councillors have direct contact with the officers. In Newbury District they are invited to, and frequently do attend, site meetings which are convened to discuss particular development proposals. These site meetings (which may also be attended by developers, local residents and other members of the public) are akin to less formal committee meetings and spectators may also present their views. Whilst the committee itself may adjourn to a corner of a site to discuss matters privately, the site meeting is the occasion closest to an open forum for debate. However, issues discussed are confined to a particular proposal and so are of limited leverage potential for the parish council. Moreover, this is a better opportunity for the parish council to use the decision making political channel rather than the recommendatory professional channel.

Direct contacts between parish councillors and planning officers are comparatively rare. County planning officers hardly ever have direct contacts whilst one district development council principal officer was able to name only three parish councillors whom he might expect to hear from on occasions. However, such contacts were usually to enable the parish councils to receive more information on particular items and seldom did they provide more information. Tensions exist because of this minimal contact, but can seldom develop, simply because of the lack of communication. As many as five of those interviewed claimed that their relations with the district planning officers was of the highest order ('excellent'). But in answering such questions, it was obvious that the councillors were generous in their assessments. The last thing most Tory councillors wanted or wished to admit to was conflict. Deference and lack of conflict are characteristics of their idealized version of government.

The parish councils which were prepared to admit to parish-planning officer tensions were those which were under development pressure and often the town councils (which we have noted tended to be much more politicized). When tensions were articulated, they were strongly stated. At times there was a simple feeling that the parish council views were given inadequate weight; others recognized conflicts of personality, but few made direct complaints about procedures.

The county planning officers felt they had good relations with the parishes and sometimes claimed their comments were more helpful than the districts' submissions. It should be noted how few their contacts with parishes are, however, and that the county is seldom involved in development control items.

From the district planning officer's viewpoint (gleaned through casual conversations), parish councils are regarded with a degree of patronage and a whisper (sometimes a shriek) of annoyance. The officers clearly welcome the wealth of local information they sometimes provide but are frustrated by their apparent lack of knowledge of planning affairs and by the political leverage they apply to their district committees.

The political channel

Whilst we have seen that parishes are given opportunities to exert influence upon planning issues through the professional channel, this is usually considered as an insufficient mode of influence and resort is often made to the political channel. The use of the political channel is generally more informal, more subtle, and believed by the parishes to be more effective. We shall first consider the links with the district councillors, and second links with the county councillors.

Block 17 Parish council observations and the district response

Figure 5.5 shows the frequency of parish council and district planning officers' agreements and disagreements on planning applications for ten parishes in Newbury District in 1978/9. The district planning committee's ultimate decision is also recorded (Pinch, 1984).

In all but one parish, the views of the parish and the officers coincided more frequently than they diverged. Since the officers may take the views of the parish into consideration in the formulation of their recommendation to committee, this high level of agreement is to be expected. In addition, the merits of many applications are readily apparent and it will be obvious to many parishes that extreme attitudes to certain proposals would be futile and decrease the credibility of the parish.

When parish and officers' views do coincide, it is nearly always the case that the district planning committee will follow their lead and issue their decision accordingly. In only six did the committee go against the jointly held views of parish and officers.

The situation becomes much more complex, however, when the views of the parish council and the officers diverge (on 35 per cent of all applications). On the one hand the local district councillor(s) may have been lobbied by the parish concerned to uphold its views, while all of the councillors will have felt the weight of the planning officers' views. In general, the view of the officers prevails with the committee, but in a few cases (17 per cent) the committee will uphold the parish view. The success of a parish in opposing the officers' recommendation varies considerably from parish to parish: whilst Purley and Stratfield Mortimer never successfully overturned an officers' recommendation, Pangbourne was remarkably successful in opposing officers' recommendations. We attributed the success of Pangbourne parish to the particular influence of their district councillor.

When the parishes and the officers disagree, there is a tendency for the parish council to want to 'block' the proposal (i.e. refuse it against officers' advice to approve). None the less, there are a substantial number of occasions (19 per cent of all applications) when the parish wishes to 'facilitate' a proposal (i.e. approve it against officers' advice to refuse). Blocking is most likely on the larger proposals whilst facilitating is most common on small proposals such as residential extensions. Overall, parishes appear no more successful in blocking than in facilitating proposals.

To ensure that a parish council has its views made known to the political component of the district council, it is necessary for the parish to establish contact with district councillors. Invariably, such contact is almost exclusively with the representative for their own area and resort is seldom made to other district councillors.

One of the most direct ways of ensuring that the parish view is heard at the district level is for the parish council to have a dual representative – a councillor who holds both a parish and a district seat. By such means the potential communication gap between

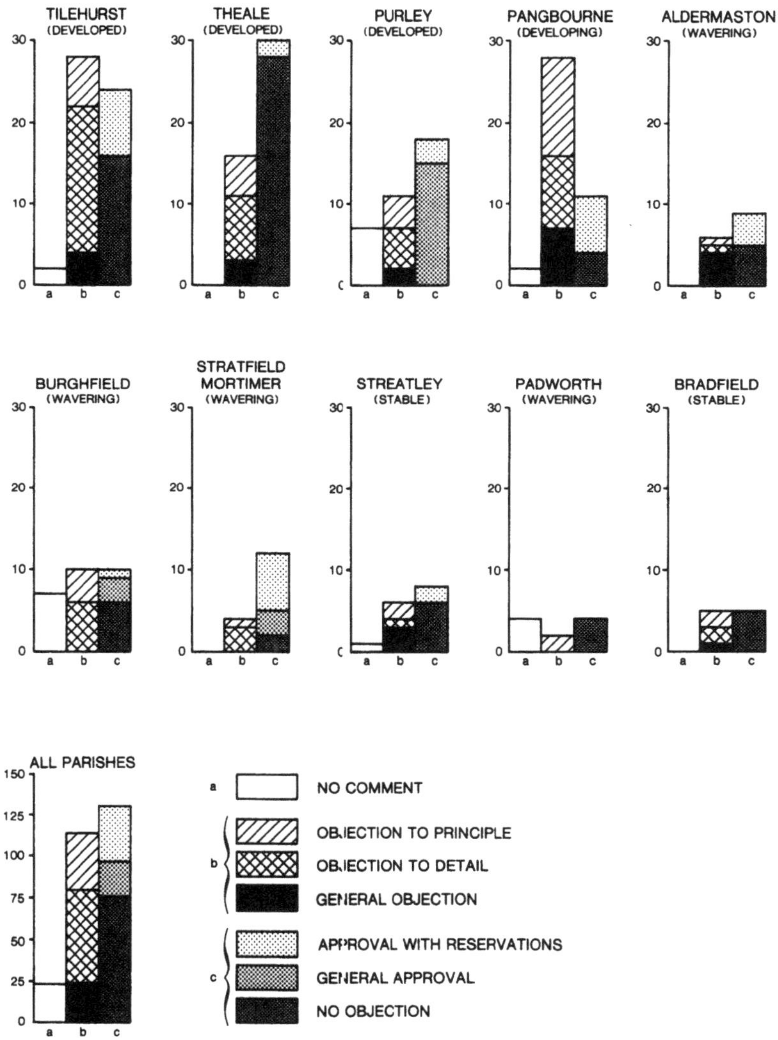

FIGURE 5.5 *Parish council observations and district council decisions in Newbury (East) 1978–9*

parish councillors and district councillors should be bridged. There was a substantial overlap of parish and district councillors: twenty of the forty-four parish councils of Central Berkshire have at least one dual representative. Furthermore, the place-focus of planning encouraged as many as fourteen parish councils to have planning dual representatives (i.e. their dual representatives held seats on the district planning committee). Of the twelve parish councillors whom we interviewed, two were dual representatives and another four were from councils which had dual representatives. This left a further six with no direct access to the political component of the district council.

Is the use of a dual representative the most effective way of ensuring that the parish council's views are heard at the district? The evidence is conflicting.

Dual representatives feel the pressures of representing two overlapping electorates at different levels of government. There will be expectations at both levels of government that the dual representative will carry weight with the other. In planning terms, the status of parish government as an advisory rather than a decision making body removes certain shackles of responsibility and allows it (or forces it) to act more in the role of a pressure group than as a tier of government. In contrast, the district council has the responsibilities of power rather than the flexibility of pressure. In serving both these tiers of government, the dual representative has to steer some middle course.

For the planning dual representatives, that middle course *usually* meant acting as a restraining influence upon the parish council when considering planning matters. Through their contact with planning officers, planning dual representatives bore the heavy weight of having to adhere to planning codes and yet satisfy (the often planning-naive) parish councillors whose frequent aim was to preserve the parish from development. It was interesting to note, however, that the non-planning dual representatives felt these pressures less when considering planning issues and were often to be seen taking a much more parish-orientated view.

The two dual representatives (neither of whom were on the district planning committee) thought that their dual role was helpful to all concerned and identified no major difficulties in performing the two roles simultaneously. However, of the four parish councillors interviewed whose councils had dual representatives (not the interviewee) a less conflict-free situation was perceived. Only one claimed to be entirely satisfied that parish interests were upheld at the district council; the remaining three had reservations. Of these three, the reservations of two concerned planning issues. Two contrasting pictures emerged.

There were occasions on which one parish councillor thought that their dual representative was too protective of the parish: 'our parish/district representative has a reputation for being almost fanatical in defence of the parish and perhaps pushes for more than we would.' This was a rather unusual case in which the planning dual representative, far from acting as a restraining influence upon the parish council, was often instrumental in pushing an even more pro-parish line. An example was quoted in which the parish council was content to have a site consist of a mixed dwelling development but the dual representative was pushing for a development consisting of all bungalows (thought to be less detrimental to the environment). Clearly this particular planning dual representative was single-minded in her defence of what she saw as parish interests and was less susceptible to the pressures of the officers than perhaps even the parish council itself.

In contrast, the second council provided quite a different picture where the dual representative was most definitely feeling the weight of the district planning officer's arguments. The interviewee claimed that the dual representative had changed her voting intentions between parish and district meetings. In a rather charged atmosphere, the parish received a slice of the 'Heseltown' allocation, an issue which was not neglected in the local electioneering which was to follow. The dual representative had failed to steer an acceptable middle course between parish and district but the issues at stake were complex and clearly the parish were not privy to all the arguments. The situation reached such an impasse that at the following election, the dual representative, the other district councillor and the parish interviewee all resigned their seats. Despite the presence of a dual representative, the parish–district communication gap was, in this instance, very large.

On parish councils on which there was no district council representative, the potential parish–district communication gap and scope for conflict is large. However, since many of these parishes were rural (and often of the stable type) and since they felt little development pressure (and indeed few other threats to the status quo), parish–district councillor relations were apparently very satisfactory to all concerned. Whereas the level of contact between dual representatives and their parish council is of necessity high, the contact between district councillors and the appropriate council will be much more variable.

Generally, district councillors are invited by the parish to attend at least monthly meetings. These invitations were accepted in four of the six parishes, showing the degree of close parish–district liaison even without the presence of a dual representative.

On only one council did there appear to be minimal parish–district contacts. This was a town council and it was clear that the district members considered that the lowest tier of local government was little more than a nuisance. Notably, the town councillors complained bitterly of their district councillors and considered that their electorate's interests were overlooked. It is difficult to know how long such disagreements had persisted but central to the issue was the fact that the district councillors were Conservatives, and the town council was Labour-controlled.

Contrasting this case was a rural (wavering) parish in which there was weekly and close contact between parish and district councillors and it was apparent that there was a tight nexus of cooperation amongst the parish planning chairman, the local district councillor and the local county councillor.

In general, those parishes without dual representatives were satisfied with parish–district contacts. Some parish councillors did not approve of the idea of dual representation and felt that parish–district relations were better if the roles were separated.

Ironically, a picture of parish/district councillor relationships emerges which suggests that conflict is most likely where there are dual representatives. However, this probably reflects the fact that dual representation is most common where parish/district conflicts are most likely: to reduce such conflict the dual representative has emerged. Resolution of the conflict has not always followed.

Parish council links with the councillors of the county

Parish-county councillor contacts were much less frequent than parish-district contacts. While invitations might regularly or occasionally be made to the county councillors to attend the monthly parish meeting, few in our sample availed themselves of the opportunity. Only three parish councils contacted could claim that they had any form of contact with their county councillors on a monthly basis and only five county councillors held parish seats. None of our surveyed parishes had double representatives (although one had an approximation of such through a spouse linkage).

Despite the general low level of parish–county communication, only two parishes were in any way angered by the situation. As one councillor said, 'I'm rather glad they don't get too involved,' but it was noticeable that this councillor's parish was in well-defended Green Belt territory: the planning system itself was sufficient defence of the parish.

Not surprisingly, it was the two councillors interviewed whose parishes were to receive substantial parts of the Heseltown allocation who articulated most concern about their county councillor. Both had requested that their county councillor take a greater interest in parish affairs (and not just in planning matters) but they considered that the response was absent or inadequate. One of these parish councillors claimed that although the county member had attended all parish council meetings four years previously, they had had no contact with her for eighteen months. Relations had soured over the issue of the closure of a local school – though the parish councillor appeared to accept the general Conservative principle concerning school closures, he felt that their village school was a special case. The county councillor did nothing to help and the parish council then turned to other political parties at the county council for support. Later came the Heseltown issue, but by this stage relations were poor and only one meeting was held at which it was considered; nothing of significance transpired. The bitterness of this parish council was shared by the adjacent parish who also felt that the county councillor did nothing to promote parish interests.

The other aggrieved parish council had not reached quite such an impasse with their county councillor. None the less, most of their communication appeared to be through written reports. Moreover, there was the suggestion that the county member showed an imbalance in the attention which she paid to the parish of her residence rather than to her constituency as a whole. An informal talk with the county councillor did little to suggest that the aggrieved parish councillor was incorrect in his assessment of the county councillor's priorities.

Parish councils, therefore, do not merely use the professional channel (comments directed at planning officers) in making their representations about planning issues. Either by incorporation of district or county members into their council or by close liaison with them, many parish councils ensure that their views are taken into consideration not only in the formulation of the recommendation by the officers but also by the ward member(s) in their issuing of a decision at committee. Probably because of their greater simplicity and specificity, development control matters usually engender the greatest interest and response from parish councils. In all but a few cases they regularly and systematically made their comments to the local planning authority. In many cases, and usually when strong feelings are evident, the district councillor is requested to further expound the parish council's views at committee. On occasion, however, parish council views may be naive or considered unduly restrictive or permissive and the planning officer's recommendation may be contrary to the parish council view. These are the instances when the district ward councillor(s) play a crucial role. Should they agree with the parish council or should they feel under serious personal or political pressure to voice their views, they will take the opportunity at committee (and before) to try to enlist the support of their fellow members. Interesting voting patterns often emerge which distinguish policy upholders (those generally siding with the officers) from the place agents (those who by claiming to judge every application on its merits, will side with the parish through conviction or as a result of political pressure). There are of course occasions on which the district member will go through the motions of supporting the parish council knowing full well the cause is a lost one: said one despairing member at committee, 'well, this is what the parish council told me to say.' On other occasions members have more bluntly said, 'I really don't know what was going through the minds of the parish council. I don't understand their view at all.'

It would take a longitudinal study to see whether the attitudes of parish councils to planning issues are changing. The affairs of Central Berkshire have certainly raised the planning consciousness of a few parish councils but most are slow to relinquish their traditional views. They persist in commenting upon development control matters – which are in many cases effectively decided by policy. Their comments on planning applications might be more useful and meaningful were they to suggest viable modifications but too often their demands are unrealistic. Where the parish councils usually miss out is in contribution to policy – that is where they could usefully convey their knowledge of their locality to the benefit of the inhabitants.

6 Residents' associations

'A lot of work. Some disappointment. A little fun.'

In this chapter we will focus attention on those voluntary organizations representing residents' interests in relatively small areas. These organizations, which are explicitly place political rather than party political, have been variously described as residents' groups, amenity societies and neighbourhood associations.

Since the early 1960s there has been a dramatic increase in the number of such groups. The annual rate of group formation, rising steadily in the 1960s, reaching a peak in the early 1970s and then tailing off in the 1980s, is reflective of the more general degree of interest in the environment (Sandbach, 1980). However, the cumulative effect as shown in Figure 6.1 is that there are now a large number of neighbourhood groups and civic societies operating in contemporary Britain.

There are many reasons behind this growth, some of which has been dealt with in some detail by Short (1984). In the following pages we will only summarize this larger exposition.

Tenure shifts in the housing market are an important background factor. The rise of owner-occupation has involved a majority of households; there are now almost 55 per cent of households in Britain in owner-occupation, a form of housing investment whose return is based partly on the quality of the dwelling and partly on the nature of the surrounding environment. Much resident group activity can be seen as an attempt to maintain and improve property prices. In the local authority sector the production of large estates has provided the environmental context for potential tenant mobilization while the steady relative increase in council rents and increasing problems of maintenance have provided important foci for such mobilization.

In terms of general political action there has been pressure from above and stimulation from below. We have already seen in

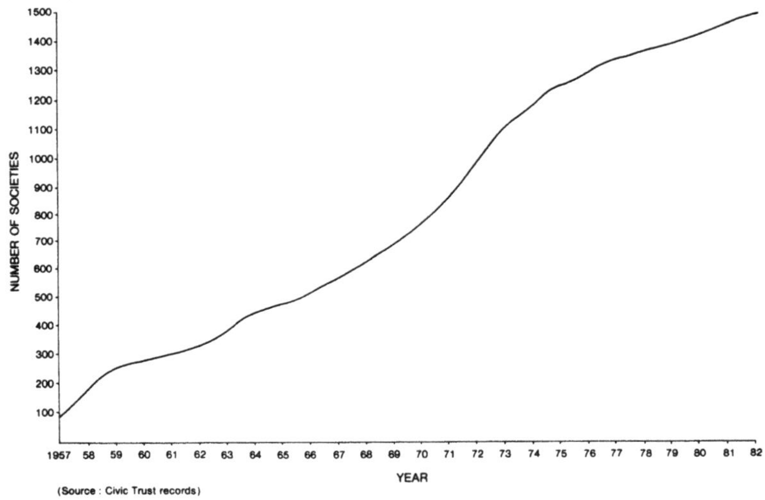

FIGURE 6.1 *The cumulative national growth of amenity societies*

Chapter 2 the general drift towards public participation in land use planning. While these top-down initiatives provide the basis for political purchase there have also been two key bottom-up (or middle-up to be more precise) processes at work. On the one hand there has been the increase in middle-income groups who have not been debilitated by the sense of inevitability, deference or resignation which has operated in the past to vitiate much local action. On the other hand many people have felt that the former political channels, dominated as they are by workplace and class considerations, have failed to adequately represent the interests of the living place and community concerns. The perceived failure of the formal political channels to articulate place-specific issues has been an important generator of resident group formation.

All of these trends have taken place within the context in which the role of the state has increased. Although there have been more recent attempts at privatization, in the long term there has been a steady increase in the level and range of public goods and services. This means that what was once the preserve of market calculations is now increasingly the concern of public negotiation and compromise. The land use planning system, for example, politicizes environmental concerns, fusing private interests, local concerns and public issues.

The rise of neighbourhood groups has not gone unrecorded and there is now a large body of literature varying from the historical (Castells, 1983) to the general (Kirby, 1982) and from the broad survey (Short, 1984) to the specific (Wates, 1976). In this chapter we will look at the rise, the form and functioning of such groups in Central Berkshire.

As we have already seen in previous chapters, Central Berkshire has been subject to successive waves of development pressure. This has involved pressure from developers and builders to construct offices, factories and dwellings. The threat of this pressure and its successful execution has provided the basis for local groups seeking to resist development. This articulated response has been fought around such slogans as *no growth*, *no growth here*, *not that type of growth here*, or *not this type of growth here now*. Not all the pressure has been resisted and new housing estates have been built where residents require and demand a range of public services. The failure of either the market or the state to provide and improve socio-cultural facilities on these new estates has been an important reason behind the development of residents' groups.

The residents of Central Berkshire, like those of other parts of the country, have also been affected by the range of externalities occurring in most urban regions. Facilities generating traffic, noise or social effects are being located and relocated all the time, throwing an externality field around surrounding areas. These changes and the more specific pressures related to Central Berkshire as a growth area are the context of resident group activity.

A total of 149 residents groups were identified in Central Berkshire. With a population of 127,722 in 1980 this gives an approximate figure of 1.17 resident groups per 1,000 population. The precise figure will be higher since it is unlikely that we recorded all groups currently active. It is difficult to place this figure in any comparative context since figures are not available for other areas in this form. Only subsequent studies will be able to say whether this is a high, low or simply an average figure.

From this total, questionnaires were successfully administered to ninety-two residents' associations who were concerned with very specific localities in the region ranging from a cluster of dwellings in one street, to a large housing estate, to, at most, a small village. The rest of this chapter will be concerned with the analysis of the responses to the questionnaire survey conducted over the period 1981 to 1983 (see Appendix 5).

Group formation

The pattern of group formation by year is shown in Figure 6.2. As with the national scene, the evidence suggests formation peaking in the mid-1970s after a steady rise in the 1960s. This

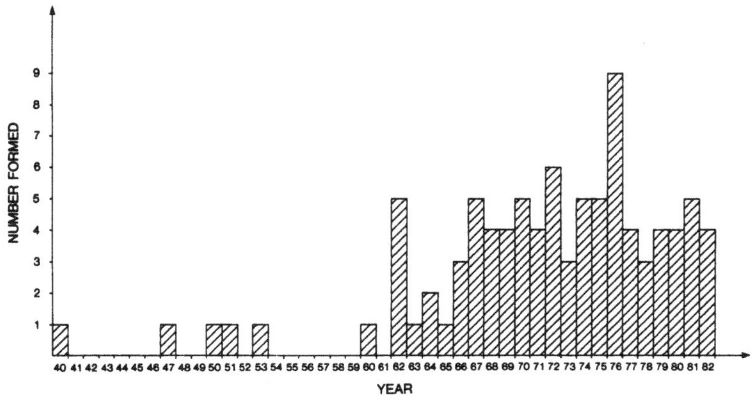

FIGURE 6.2 *Residents' groups in Central Berkshire*

picture is consistent with the trends in housing tenure, the encouragement of public participation and the rise of an articulate middle-income group in society which we have already noted. The 1960s and 1970s saw a steady increase in owner-occupation, increasing use of participation schemes in land use planning and the emergence of a white-collar workforce. The formation of residents' associations reflects these trends while also being part of the course of official recognition of the need to placate, involve and defuse community group pressure.

The cumulative figure, presented in Figure 6.3, shows a steady, inexorable rise, with increase in the number of groups recorded each year since 1960. The data in Figure 6.3 are presented to distinguish between active, dormant and defunct residents' groups. Active groups (n = 76) are those whose internal organization is working to achieve certain goals and objectives. Dormant groups (n = 12) are those whose organizational structure exists but is not being used currently to achieve any goal. Six defunct groups were identified and interviews were held with representatives of four of them. A defunct group is one which no longer exists, whose organization has been disbanded and has no effective role in the local area.

It is possible to tentatively outline the *life cycle* of a typical residents' group. It appears that a group forms around a specific issue initiated by dynamic individuals utilizing local resources to work against a perceived threat and/or to meet a perceived unfulfilled need. When the issue has been resolved either to the group's advantages or otherwise the group may then be disbanded or become dormant. This is most likely to occur when the group

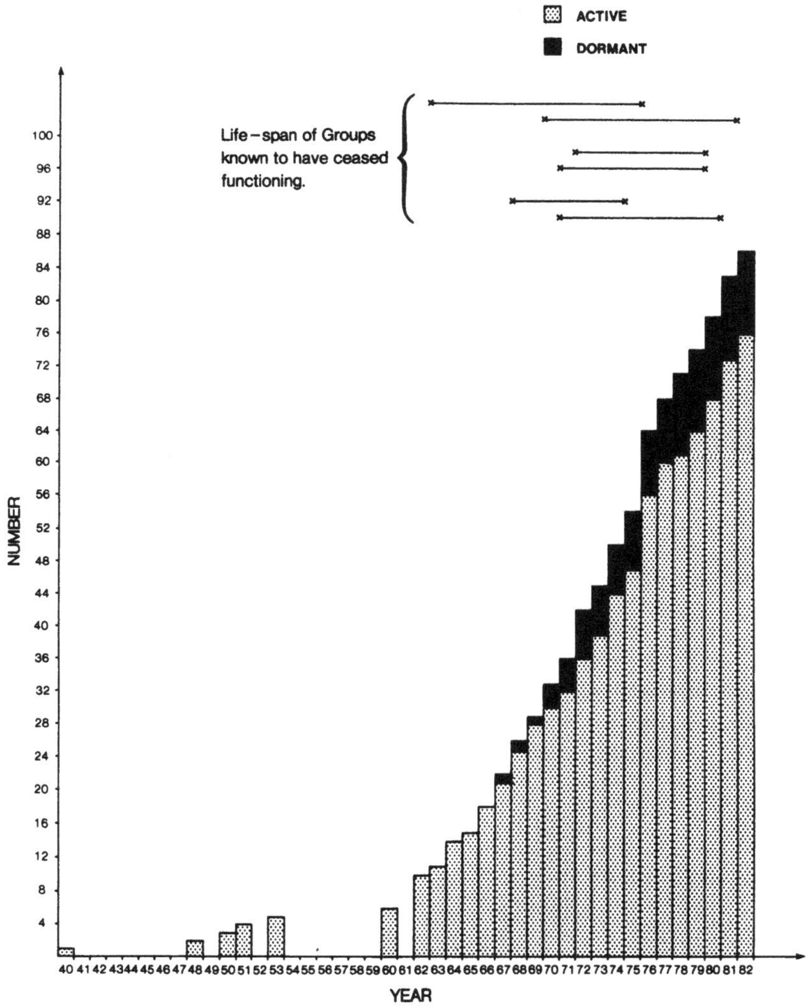

FIGURE 6.3 *Residents' groups in Central Berkshire: the cumulative picture*

concentrates its efforts on one specific issue. Alternatively the group may continue because the objectives have widened and broadened (e.g. a residents' group established to fight against a new road may continue in existence to 'preserve the character of the area', establish a social centre and/or to keep residents informed of local planning matters) or because similar sorts of threats and needs continue to appear. The cumulative figure shown in Figure 6.3 is thus likely to conceal a number of groups which had only a brief existence, brought to life and extinguished by ephemeral events. However, the large number of active groups indicates that in Central Berkshire residents' groups either may have broadened their aims, or the reasons for their emergence continue to make groups active.

The why of group formation

All resident groups are concerned with their local environment. The term 'environment', however, needs careful inspection. Robson (1982) identifies three types of environment. The *physical environment* is as the name implies the type and extent of buildings, roads, open space and all those elements which go to make up the built and natural environment. The *social environment* is the social and demographic characteristics of the local population, while the *resource environment* refers to the location, distribution and access to private and public goods and facilities such as shops, schools, recreational opportunities, etc. The different residents' groups are concerned in varying ways with different environments. From the responses to questions probing the why of group formation it became clear that groups were concerned to varying degrees with both stopping developments in the physical and social environments and obtaining things for the resource environment. These two characteristics can be summarized as 'stopping' (or protecting) and 'getting' (or enhancing). The precise emphasis varied and three broad categories were identified.

'STOPPERS' (n = 40, 43%)

These were groups primarily concerned with protecting local areas from further development. Typically, they object to some development planned for their area such as a new housing estate or a new road scheme which they believe will detract from the physical and social environment. Just over half of these groups describe their aims as stopping or modifying unwanted development while the remainder saw their main goal as protecting the quality of their environment. Development here means a major perceived change in the physical and social environment. In 83 per cent of cases group formation was initiated by a specific issue. These groups are primarily concerned with protecting the physical and social environments from further change, with the explicit vocabulary of physical environmental concern often concealing an equal if not greater concern over the social environment.

The concerns articulated in no-growth-here stances reflect both economic calculations and specific social valuations. They are economic in the sense that house prices are perceived to be at stake since these groups are predominantly owner-occupied and overwhelmingly middle-income. Growth is seen as imposing a negative externality on house prices. There is a hard core of material interest underneath the environmental concern, what Frieden (1978) has termed the environmental protection hustle. In association with the direct concern with the effects on house prices there is also a more subtle and indirect concern with the social composition of the neighbourhood in relation to further residential development in the area. The presence of lower-income settlers may affect property values. But there is an independent social evaluation. In the more salubrious urban neighbourhoods and villages many households have located because of their exclusivity. Generally, the high price of housing operates to maintain exclusivity through informal control rather than direct manipulation and this tends to feed the ideology of a harmonious 'natural community'. Particularly in the rural areas, there is a powerful ideology which sees in a village location the hope of reasserting a moral arcadia away from the anonymity of mass urban society. The concern with community here is an attempt to face up to modernity by asserting definable positions within a small local social hierarchy. The defence of villages from development mainly comes from residents of less than twenty years' standing who wish to maintain the physical village on the ground as much as the village in the mind. Not all villagers are involved; it is mainly the middle- and upper-income residents fearful of the disruption of the village hierarchy where everyone knows everyone else and a subtle social order is maintained because everyone knows the rules guiding social action and interaction. Defence is greatest when developments are proposed which either lead to the coalescence of villages or the submerging of a village by a larger town. Such developments strike not only at the material base but also at the emotional heart of the stoppers.

In Central Berkshire the stoppers' groups cover the interests of certain residents in the middle-income urban and suburban neighbourhoods as well as those in villages. Two-thirds of the stoppers' groups are located in rural Central Berkshire while the remainder are in the middle- and upper-income neighbourhoods in the major towns. We can consider more clearly the nature of these groups with reference to three specific instances.

Hurst Village Society is a typical rural pressure group whose interests centre as their name indicates around the village of Hurst: a mixture of old cottages and large detached houses built in the post-war period. The group was formed in 1972 with 150 members. The initiative for the group's formation came from a few residents protesting about gravel extraction which disrupts the physical landscape and generates heavy commercial traffic. The group sees its overall role as preserving the character of the area. The physical and social environment are intertwined in this concept of environmental protection and character maintenance. With a membership of just over 200 the Hurst Village Society is affiliated to the Council for the Preservation of Rural England (CPRE).

The Society for the Protection of Ascot and its Environs (hereafter the Ascot Society) was founded in 1971 and now has 800 members covering north and south Ascot, Sunningdale and parts of Cobham Common. This is stockbroker belt *par excellence*, the location for upper-income residents in both Central Berkshire and parts of Surrey. The Ascot Society was initiated by a group of residents concerned about the development of 100 houses in an estate development. The perceived threat was both to the physical and social environment. The broad aim of the group is to resist development pressure. As one member noted, 'Unless planning controls are exercised in accordance with the structure plans then the area will be swamped by those wishing to live here. The aim of the society is to see that concepts contained in structure plans are adhered to.' The Ascot Society in its constitution formally sets out its purposes:

(a) To endeavour to ensure that all development within Ascot and the surrounding district be desirable to the majority of the inhabitants of such area.
(b) To endeavour to ensure that all development shall be consistent with the existing character of the area.
(c) To ensure, so far as possible, that all development shall be in accordance with the policies and provisions set out in the improved Ascot Town Map of 1960 and in that part of the development planned for the County of Surrey applicable to Sunningdale, in both cases subject to such formal amendment and such planning approvals as exist at the date of the adoption of this constitution.
(d) To ensure, as far as possible, that all development shall be within the policy for Metropolitan Green Belt areas where applicable, the said policy being as stated in the oral answer given by the Minister for Local Government and Housing on the 19th May 1971 and reported in Hansard at 1247–1248. To seek to retain the existing natural features of the area, including the flora and fauna.
(e) The activities and responsibilities of the society shall not be confined exclusively to the area of Ascot and

Block 18 A stopper group member

Mrs Z is a 41-year-old housewife with two young children. She has lived in her present house for the past two years. She is a member of a stopper group.

'I joined the ... Residents' Association when we moved to Wokingham in 1973. The previous householder had been a committee member and mentioned it to us: I thought it would be a useful way of finding out what was going on in Wokingham, and knowing what sort of particular problems were concerning people in our area.

'I have always been on the committee of the association, and for two three-year periods have been the secretary. I took these commitments on because no one else volunteered. Being secretary involves taking minutes (though these are not typed up or circulated), writing some letters, writing the newsletter and arranging for it to be duplicated and delivered – my children often do the latter.

'I don't have high aspirations of personal fulfilment in the association. I have probably got to know more people living round me than I would otherwise have known. I hope it will continue as a working body, but it needs very serious issues that radically affect people before we get much response from the membership at large. We might get a more active membership if we had a more vibrant leadership (and I count myself as not being so).

'I think our greatest success was the formation of the babysitting group about six years ago. Before I came here, the association, along with others, had had great success in getting a public inquiry into the filling in of the Blue Pool site behind one of the roads in the association. Severe restrictions were placed on the way this land was treated because of the association's actions. We have failed to get any enthusiasm for social activities, and as far as I know, have never held any. The only thing we ever organized was a children's slow bicycle race for the Jubilee celebrations.

'I think our relations with local authorities have been most successful when our views have coincided with theirs, i.e. an application for housing development behind ... Avenue was opposed by us and the district council. The case went to appeal, and was turned down by the DOE Inspector (primarily on a technicality), though the fact that for years we have complained about the tendency for our gardens to become waterlogged in winter had got through to the council, as it was briefly mentioned by them in their written representations and was noted by the inspector. However, the plan about eight years ago to fill in a pond in ... Avenue and build three houses on the site, although bitterly opposed by us, went through. I don't think our arguments have ever changed council policy; what we have done is to contribute what have turned out to be majority opinions, when the Berks County Council has asked for the views of the public on matters like the inner distributor road for Wokingham (on the shelf at the moment) and the areas of Central Berkshire which should be developed under the Area 8 Strategic Plan.'

> surrounding district where such activities outside such area can reasonably be expected to reflect beneficially upon the area of benefit, i.e. Ascot and surrounding district.

Erleigh Road Residents' Association covers an area of Victorian and Edwardian housing ranging from large houses, terraced dwellings to small amounts of new high-income infill in Reading. Concern here was crystallized by middle-income residents around the attempt by landlords to provide cheap bed and breakfast accommodation just over three years ago. With a membership of 100, Erleigh Road RA is a stopper group in an inner-city context. It is not so much development pressure as externalities from changes in the area which constitute the nub of concern. Subdivision of housing, the location of half-way houses for alcholics and the problems of road safety are the typical cases concerning this residents' association primarily comprising middle-income residents in owner-occupied dwellings.

'GETTERS' (n = 34, 37%)

These are groups whose main concern is with the enhancement of the local area through the provision of community facilities and the generation of a community feeling. These groups are concerned with both the social or in this case community environment and the resource environment. Many of this general category use the term community rather than residents' association. There are two specific types. The vast majority (88 per cent) are based on recently constructed private sector housing developments. An exemplar is *Caversham Park Village Association* (CPVA), formed in 1965 when thirty households in a new estate voiced concern over community facilities in the area. CPVA now has 1,700 full members and 300 junior or associate members on a estate of 1,500 families. Almost 60 per cent of houses have at least one kind of member. The constitution defines CPVA as 'an educational charity, formed in 1965, to provide a common effort between Local Authorities and inhabitants to advance education and to provide facilities for recreation and leisure'. There are only three groups in this category covering public housing areas – Dee Road Residents' Association, Coley Park Residents' Association and Waterloo Road Residents' Association. *Dee Road RA*, for example, was formed in 1977 with the aim of fostering community spirit and improving facilities on the estate. Dee Road is a 1,000-dwelling council housing estate built in the 1960s in the period of high-rise system built council accommodation. Like many of the public sector high-rise dwellings built at this time it has severe

Block 19 A getter group member

Mr Z is an engineering manager in his forties. He is a married man with two children and has lived in his present owner-occupied house for the past ten years. He is a member of a getter community association.

Community and Residents' Association – member profile

'What were your reasons for joining the group?'
'A desire to fill some of my leisure time and a desire to contribute something useful to the community.'
'What is the nature of your involvement with the group?'
'I first joined the Community Association ten years ago and joined the Bridge Section. I found that people there spent a lot of time talking about CA activities as well as talking about bridge. I was gradually drawn into these conversations and asked if I would be prepared to do a job in the CA. I said I would and asked what jobs were still unfilled in the coming year. I was told that the only job definitely vacant was that of Centre Facilities Officer (soon to be called Development Officer). I agreed to stand and found myself with a "full-time" job to fill my spare time. The parish council had just brought five acres of amenity land from the estate developers and were prepared to lease part of it to the CA to enable them to build a community centre.

'So my first job in the CA was to coordinate the design and construction of our first building, now known as the Social Centre. I had a committee of twelve to fifteen people to help me and it involved us all in lots of hard work. Towards the completion of the building contract (about two years later) I found that there was so much organization still to be done that I applied for 'social service leave' from my company (the company runs a scheme whereby they granted paid leave for social projects). I was granted two months' leave to complete the work, and our first building went into operation in December 1976 and was officially opened by Sir Harold Wilson on 8 January 1977.

'I continued work as the Community Association Development Officer in 1978 and with my committee we completed the preliminary design work on our Youth and Recreation Centre. However any further work had to be shelved until 1981 due to lack of funds.

'In 1979 I was elected Secretary and in that role my responsibilities were for:

(1) Meeting minutes and Agenda
(2) Membership
(3) New arrivals welcome scheme
(4) Licences and all legal matters
(5) Constitution and rules of CA
(6) Standing Orders Committee
(7) Correspondence

'All this could not be done by one person and I was very lucky to have the help of other volunteers who very efficiently assisted, namely a Minute Secretary, a Membership Secretary and a New Arrivals Secretary. I was also very fortunate to be able to enlist the very latest in office technology to help me (word processors and photocopiers),

Block 19 continued

as I work for Rank Xerox and they were quite happy for all the equipment to be used outside office hours.

'In January of this year (1983) our President retired and I was elected into the post. A refreshing change after being Secretary for five years. My responsibilities are now:

(1) External liaison
(2) Planning Committee which is concerned with planning the long-term future of our association.

'On a personal level, what do you get out of belonging to this group?'
'I get a lot of personal satisfaction from being part of a very successful voluntary group. My involvement, like everyone else's, has a two-way benefit. Hopefully the CA benefits from my contribution, and I know that my skills have increased tremendously from my CA activities, e.g. the experiences of managing large building projects, the wealth of knowledge gained from learning about licences and legal matters, the problems of interacting in a large group of volunteers, all these things contribute to making me more effective in my (paid) working environment.

'In a consideration of the groups' links with other such groups and with the local authority, which issues do you personally feel have been successful and which have been problematic?'
'I do not feel that the CA has been very successful in establishing links with other local groups. We do have local groups affiliated to the CA, examples are the tennis club, the local Brownies, Aquanauts, etc.; also affiliated are the Committee of the Church, and the Women's Institute. These links are reasonably successful but we do not actively seek links with other groups and our links with other CAs are virtually non-existent. This is something I would like to change and I am at present looking at ways we could try to form a Berkshire Federation of CAs. I feel the reason for our tendency to be inward-looking is the heavy burden of maintaining the volunteer workforce which means that most people just do not have the time to look outward.'

Diary of activities for week commencing 21st March 1983

Monday	09.00–18.00	Work
	18.00–19.00	Travel time to home
	20.00–23.00	A meeting at my house. Working party to discuss the new structure we will need to meet the requirements of the Charity Commissioners regarding the operation of our bar.
Tuesday	05.30–06.00	Get my daughter up and take her to swimming training
	06.00–07.00	Fill in a bit of this questionaire before breakfast (yes really!)
	09.00–18.00	Work
	18.00–19.00	Travel time to home
	20.00–23.00	Planning meeting. The Planning Committee met with a representative from Bulmershe College to agree objective for a village survey that we will undertake with their assistance.

Block 19 continued

Wednesday	09.00–19.00	Work
	19.00–20.00	Travel time to home
	20.00–23.00	Meeting in my house of the four officers to discuss rescheduling of the workload of our part-time clerical assistant plus confidential discussions on a couple of important CA issues.
Thursday	09.00–17.30	Work
	17.30–18.30	Travel time to home
	20.00–22.00	Attend Squash Club AGM
	22.00–22.30	Meet new volunteer CA Youth Officer
Friday	09.00–17.30	Work
	17.30–18.30	Travel time to home
	18.30–23.30	Leisure time with family
Saturday	10.00–12.00	Shopping

The rest of Saturday and Sunday was leisure time with the family. This was not a typical week (it just happened to coincide with my starting to fill in this questionaire). I would say that in an average week I spend 10–12 hours on related activities.

problems of dampness and condensation because of faults in the building process. The area now has a bad reputation and is used by the council implicitly rather than explicitly as a dumping ground for problem families.

These enhancers' organizations are in the business of getting things. Primarily their concern is to improve the quality of life in the local area either by pressing for improvements to their physical surroundings or by attempting to encourage community life and hence self-help through the provision of community centres and the organization of social events. They are also concerned with stopping or modifying what they see as deleterious development but their main aim is the provision of community facilities and events.

'STOPPERS-GETTERS' (n = 18, 20%)

These are groups concerned in equal measure with both the protection and the enhancement of their local area. They include old established urban neighbourhoods as well as new estates. In only one case does the area constitute a predominantly local authority housing area, and in only three cases is it mixed public and private. In the remaining fourteen the predominant tenure type is owner-occupied dwellings.

Katesgrove is an inner-urban neighbourhood in Reading consisting of late Victorian and Edwardian brick-built terraced and semi-detached housing. In Reading's housing market it is the lower middle end of the urban housing market, not inner-city but neither salubrious suburban. *Katesgrove RA* was established in 1971 and now has a membership of 500. It started primarily as a response to the plans for an inner distribution road and the consequent threat to property values. One particular individual had his house compulsorily purchased but fought the case successfully and this was the focal point for about forty people who began to get the RA established. The aims are now broader and as a broadsheet of Katesgrove Residents' Association distributed in 1982, headed *People First*, noted,

> We campaign to ensure that we get our fair share of the cake. We meet and lobby our MP, councillors and council officers. Whether it is new housing, pelican crossings or leisure facilities, we get things done.
>
> We campaign, petition, demonstrate to protect ourselves and this area; to reduce lead pollution, to cut down traffic; to reduce industrial nuisance.
>
> We keep people informed through public meetings, through the press, radio and television; and through the local magazine *The Katesgrove Clarion*.
>
> We help to organize entertainments, discos; there is even a carnival being planned.
>
> We keep in touch with all the other groups in this area and community organizations in the whole of Reading.
>
> Above all we respond to your requests for help.

Fords Farm, in contrast, is a 1,000-dwelling private sector estate in the middle and upper range of the market to the south of Reading. The lowest-priced house in 1979/80 was £30,000 and the highest was £65,000. The estate was constructed throughout the 1970s and *Fords Farm Residents' Association* was formed initially to fight for play areas and fight legal cases against the contractor for damages for flood damage. (Fords Farm is built close to the Kennet flood plain.) Approximately 200 joined the group at the outset in 1978 and the membership now numbers 800. The aims have widened to include not only the attempt to obtain public facilities such as post offices but also to restrict further private and public sector housing developments.

We will continue to use these case studies throughout the remainder of this chapter as an example of the different types of groups and their organization. We will use them as pegs on which to hang the more aggregate analysis.

Block 20 A stopper-getter group member

Mr Y represents a stopper-getter group in the eastern part of our study area. He is a 39-year-old married man with two children. He works at the University of Reading. He has lived in the same house for fifteen years and his home is now valued at between £50,000 and £70,000.

'The group I am going to talk about here is the ... Residents' Association. The idea for this group came from some of my neighbours and I was appointed chairman from the outset in 1978. The reasons for founding the group were that for some time we seemed to have been harrassed by the county council and the issue came to a head over parking, where the county council were trying to stop parking in quite a large part of our area. This was despite our own representations as individuals and those of the parish and district councils who were on our side. The main point of our group is that it is to represent the views of local people to the press and local authorities and to make sure that our problems are known about.

'I have been chairman of the association since it started in 1978 and I suppose I continue in this position largely because there is no one else who could take on the role or at least has my local contacts, that is with the local council and with a local amenity group. I also have the right technical qualifications for being in this position, having a working knowledge of transport and planning matters. You have not actually asked for information about how the group is organized but it is organized on the basis of area representatives who are one person from a street or part of street perhaps containing forty or so houses. We have quite a large membership of over 300 and our committee structure seems to work quite well.

'On a personal level, I suppose I am motivated by a feeling that as one of the few (in planning and transport terms) technical people in the area, I feel somehow that I have a responsibility to do what I can for the area. The group has been quite effective as the result of a certain amount of cooperation with local political people in getting its view across and now has quite a (notorious) reputation in the area as at least sticking up for what's required locally. One of the problems with ceasing to be chairman is to find someone else who will carry on and will ensure a level of continuity.

'One internal difficulty in the group has actually been holding a committee of twelve or thirteen people together and resolving the personal conflicts which arise all too easily. We seem to have a fair number of "idiosyncratic" members and we have had our rough times in ensuring that all points of view are heard and as far as possible taken into account. Touch wood, we are reasonably stable as a group and the various warring factions are reconciled to each other's existence. One of the obvious and big problems is getting people to do things whether it is writing things for public inquiries or picking up litter or sticking things through letter boxes. It always seems that the busy people are the only people you can actually get to do anything. This issue of delegation is particularly hard and difficult in groups of this type, especially as they do need to operate continuously and not just in fits and starts.

'Our group is one of the few residents' associations in the area and seems to be one of the better known and more successful ones. We

Block 20 continued

do not have any links with other residents' associations although we have fairly close links with the town's Amenity Society. We are fortunate locally in that our residents' association is non-political although three of our members are very actively involved in local politics. This helps because we are at least seen as not beating any political drum but just looking after our own local interests. Indeed, all political interest is buried in the cause of getting what we need locally.'

Local activists have to fit in their activities in their spare time. The core members of groups have extra strains placed on their time. When asked to note the contents of a typical week this respondent replied in the following manner:

'I have chosen a recent week for you and I work full-time, 9 to 5 in Reading, meaning that I do not usually get home until 7 or so. In one fairly recent typical week – that is the week beginning 14 February [1983] – my evening activities were:

Monday	Planning and General Purposes Committee Meeting at Town Council
Tuesday	Residents' Association Committee Meeting
Wednesday	Amenity Society Committee Meeting
Thursday	Chiltern Society Road and Transport Committee Meeting
Friday	Labour Party Group Meeting
Saturday	Looking after children while wife worked.'

The life cycle of residents' associations

We have already noted a few pages previously how residents' associations have a definite life cycle as they progress from an active to a defunct status. A similar tale can be told with reference to the aims of the groups. The classification adopted labels groups at only one point in time. However the aims of groups may broaden or change through time. Figure 6.4 presents in diagrammatic form a series of possible life cycles. In the case of (1) perceived developments create the precondition for the emergence of a stopper group. A typical case might be the threat of a new lower-income housing estate. In the case of (3) a getter group is formed, as in the case of Caversham Park Village Association in its early years. A more unlikely case is (2) where a stopper-getter group is formed. In the cases of (4), (5) and (6) the groups may then become dormant as circumstances change, either being reactivated later or becoming defunct. In the cases of (7), (8), (9) and (10), by contrast, the aims of the groups may change. The paths of (7) and (8) indicate groups which become more concerned with widening their objectives, (9) and (10) indicate narrowing objectives while (11) and (12) suggest changing objectives. In

Block 21 Changing aims and roles of a residents' group

Mr X is a key member of a residents' group classified as stopper-getter in our typology. He has lived in a newish estate for the past four years. He is a married man in his fifties with two children. He is an air traffic control officer and has a mortgage on a house valued by him in 1983 at between £40,000 and £45,000.

'What were your reasons for joining the group?'

'In common with most residents at the time, I was horrified at the prospect of a double-decker bus passing within a few feet of my front door. Also at this time the general living conditions were very poor due to a thoughtless developer who thought that no one would stand up to him. Our group did! Another bone of contention was the appalling lack of after-sales maintenance experienced by the residents – some with quite serious faults. This could also be traced back directly to the developer. It was clear that a united approach was needed to solve these problems and consequently I didn't hesitate to become a founder member.'

He quickly became honorary secretary and in 1983/4 was deputy chairman.

'What do you get out of it?'

'A lot of hard work! I don't think I aspire to local politics or anything of that nature, but I do react against the traditional view that you can't beat city hall. You can if you try and your arguments are reasonable. We've proved it.'

'And the future?'

'The association has now reached a point in time where there are immense pressures being brought to bear to change our image and our function. It worried me that these changes could be forced upon us prematurely. The traditional role of the RA representing the residents of a particular estate is clearly limited, and the time will undoubtably come when we must become less parochial. When the last bricklayer has left and the last roadway has been handed over to the local authority then the prime function for the group is to protest about dogs fouling the footpath and whether we are getting value for money from our rates, then a large, less vocal if not less active association will be indicated.'

Central Berkshire the predominant paths have been (1), (3), (1) → (7), (3) → (8) and (3) → (12).

The distribution of residents' groups

The dominant trend in the distribution of the general sample is for a concentration around the major urban centres of Reading, Bracknell and Wokingham, with a more varied, thinner spread in the rural areas. Figure 6.5 takes population density differences into account by depicting residents groups per 1,000 population by parish. Although there is a general trend for little apparent

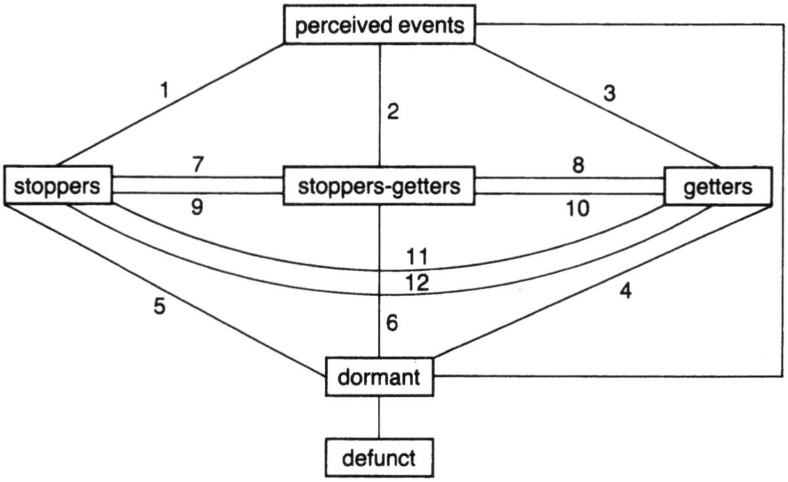

FIGURE 6.4: *Life cycle of residents' groups*
Note: See text for a discussion of the numbers.

activity in the western part of the district compared to the east, reflecting the general nature of development pressure in this area, the resulting pattern is still a very rich mosaic which is relatively difficult to interpret. However, we can use the categorization of parishes used in Chapter 5 to aid the interpretation. Table 6.1 shows the variation in residents' group activity by the four different parish types. The parish types obviously vary in population size from a 'stable' parish such as Englefield with a population of 169 to a 'developed' parish such as Earley with a population of 11,666. But these differences in population size, according to Table 6.1, do not seem to appear to influence residents' group activity. The index for all but one of the parish types are very similar. The obvious exception is the 'wavering' group of parishes where there is evidence of a much higher level of resi-

TABLE 6.1: Residents' group distribution by parish type

Parish type	*Average population size*	*Average number of groups per 1,000 residents*
Stable	555	0.36
Wavering	3,014	0.52
Developing	3,550	0.33
Developed	9,432	0.32

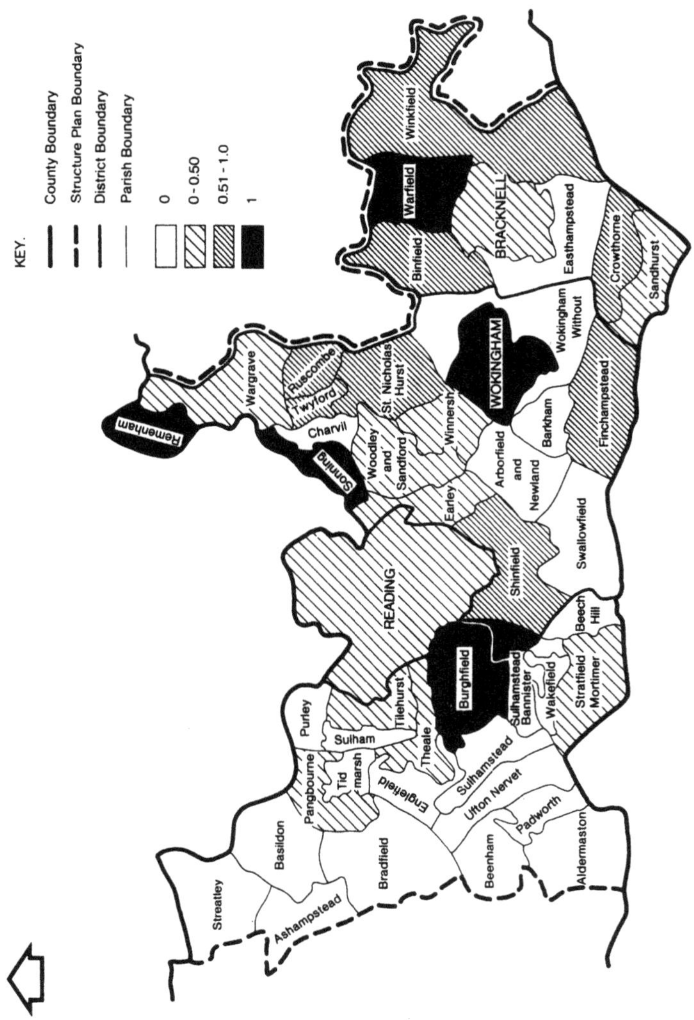

FIGURE 6.5 *Residents' groups per 1,000 population by parish*

dents' group activity. Possible reasons include the perception of possible developments providing a local climate of opinion which allows easier formation of residents' groups.

Residents' associations have a definite territorial base. Figures 6.6 and 6.7 respectively shows the distribution of the Reading- and Wokingham-based groups. Notice how there are very few instances where territories overlap, the main boundary markers being major roads. Two broad types of territory can be identified. These are the small, very specifically based groups such as Marlborough Avenue Residents' Group (no. 24 in Figure 6.6) and Chestnuts Residents' Association (no. 4 in Figure 6.7), and the much larger groups such as Caversham and District Residents' Association (no. 2 in Figure 6.6) and California Residents' Association (no. 18 in Figure 6.7). Not all the area of the two towns is covered by residents' association territories. The pattern is more like a patchwork quilt of residents' groups operating in some areas because either social composition or the stakes or the external pressures are high enough. However, it does seem that substantial portions of social space have some form of residents' group representation.

Internal resources

The membership

Events do not simply call forth residents' groups out of thin air; their formation requires the efforts of individuals. The basic resource of residents' groups is the membership which provides not only money and peoplepower (the term manpower is singularly inappropriate in this context) but also respectability. Residents' groups, in order to deal effectively with formal institutions, must be seen as representing local concerns. A wide membership provides both the capacity and the legitimacy for effective lobbying.

For all residents' groups a distinction can be drawn between a core group of active members who shoulder the bulk of organization and administrative duties and a much larger periphery of less active members who share the aims if not the enthusiasm and commitment of the core. We obtained information on core members of twenty-one separate groups in a follow-up to the main survey (see Appendix 5). The variation in response did not allow for a precise quantitative approach but some indication of the replies are given in the blocks in this chapter. However, on the basis of the replies and the actual interview it is possible to make

Key to associations

1 Mapledurham Area RA
2 Caversham and District RA
3 South Earley RA
4 Burcomeb Way RA
5 Caversham Park Village Association
6 Warren and District RA
7 Caversham CA
8 E.P. Collier CA
9 Lower Caversham RA
10 Norcot and Lower Tilehurst RA
11 Great Knollys and Central RA
12 Friends of Curtis Wood
13 Fords Farm RA
14 Southcote and Manor Farm RA
15 Coley Park Community Association
16 St Giles RA
17 Katesgrove RA
18 South Reading CA
19 Whitley Wood CA
20 St John's RA
21 Newtown CA
22 Erleigh Road RA
23 New Road RA
24 Marlborough Avenue and Elmhurst Road RA
25 Northcourt Avenue Ratepayers' Association
26 Shinfield Rise RA
27 Loddon Earley RA
28 Watmore Lane Group
29 M4 Linkway Association
30 South Lake RA
31 Maiden Erleigh RA
32 Coleman's Moor Association
33 Sandford RA
34 North Woodley RA
35 West Drive RA
36 Dee Road RA

RA = Residents' Association
CA = Community Association

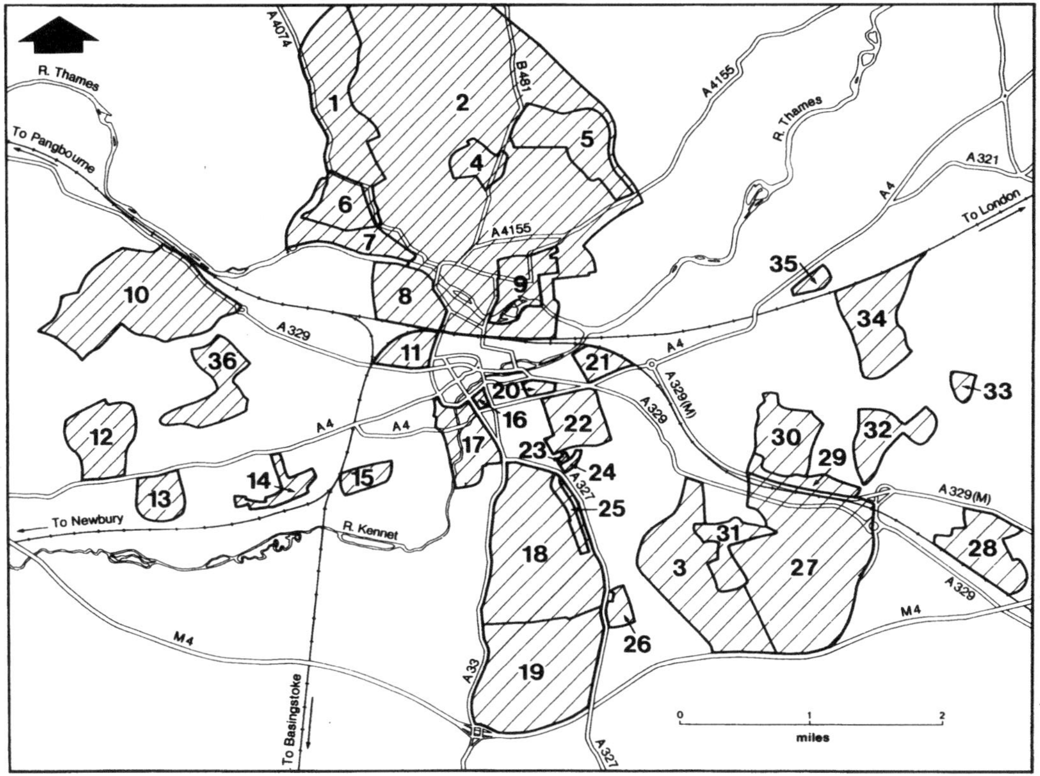

FIGURE 6.6 *Residents' groups in Reading*

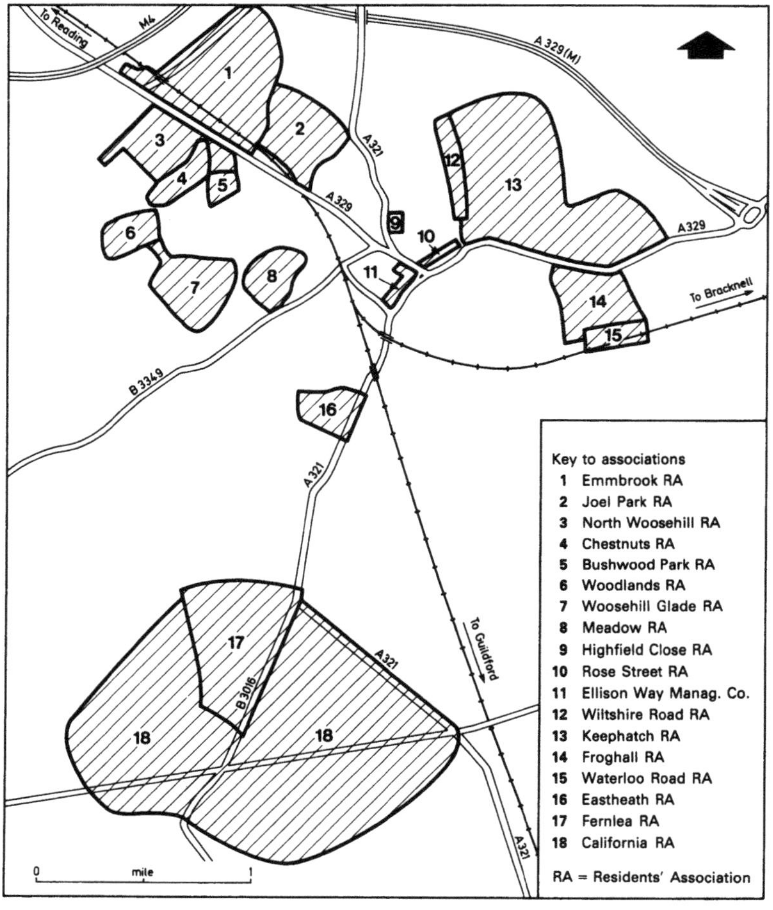

FIGURE 6.7 *Residents' groups in Wokingham*

some generalizations. Core members have four different sets of reasons for becoming involved, with the precise mix varying between individuals. A basic reason is *to meet people*. This form of social interaction is valued by those simply seeking human contact and those seeking status affirmation outside of their work and family. As one person put it, 'My main satisfaction is pleasure from helping people.' 'Getting out and meeting people' was an important reason for people getting involved with local voluntary groups. People are also motivated by a *sense of duty*. Such people either have relevant expertise which they believe may be of some use and/or they acutely perceive the need for 'something to be done'. As one high-minded individual noted, 'I hope my involvement leads to a better living environment for all residents.' One respondent noted that his main satisfaction was 'a sense of duty performed. I hope and believe all residents should take a stint on the association.' People also become involved as a way *of finding out about local affairs*. 'Since becoming a member of the association,' said one core member, 'I have learned a great deal about the town, I feel that I have put down strong roots in the place. It is good to be able to talk to traders, local government officers and councillors.' For some people involvement in a local residents' group is a quick entry into local life as much as a cause, and a means of increasing their local standing and knowledge. Finally, people may also become involved as a way of *influencing events*. A forthright activist wrote, 'I do react against the traditional view that you can't beat city hall. You can if you try and your arguments are reasonable. We've proved it.' The relatively low citation of this particular reason is mainly a function of the fact that such movers and shakers are more likely to see greater opportunities in local and central government or in the realm of work. In reality a whole combination of overlapping reasons are likely to make people join groups and become active.

As can be seen from the blocks, residents' group core members have to devote considerable amounts of their limited spare time to running the groups. There is a disparity between the part-time voluntary workers of the community action lobby seeking to do things in their spare time and the amount of time available to the full-time professional staff of local planning authorities and builder-developers. There is an asymmetrical resource endowment resulting from this disparity which commitment and dedication may not be successful in bridging.

From the nineteen usable replies to the survey a number of generalizations can be made about core members. In terms of occupation the respondents were either housewives (21 per cent) or white-collar workers, with the largest single group (47 per cent) being classified as in either professional or managerial categories. No manual workers were represented. There seems to be a definite bimodal age structure consisting of young married people in their thirties with children under 10 and another group of married people in their late forties and fifties with grown-up children. The former age category is most strongly represented in the getters while the latter is the dominant group in the stoppers. Over 80 per cent of the respondents were owner-occupiers and only a quarter of these had dwellings with current market values of less than £35,000. The modal market value range was between £35,000 and £50,000. As a group, core members are of a high status, more affluent and longer settled in comparison to the rest of the population. A thumbnail sketch would present them as middle-income owner-occupiers with families.

Residents' groups seek a wide membership base in order to mobilize resources and to secure their legitimacy when facing external bodies. Elected representatives on both district and county councils are particularly sensitive to the issue of representativeness. Aware of the danger of being written off as unrepresentative and mindful of the need to obtain voluntary support, all residents' groups seek to widen their membership base. The size distribution of membership is shown in Table 6.2. Notice how the majority of groups have only a small membership reflecting the circumscribed area of concern. Residents' groups by their very nature are parochial. In total, however, the residents' groups have a combined membership of 44,000 which represents almost 12 per cent of the total population of Central Berkshire. In other words more than one out of every ten people in Central Berkshire belong to some form of resident association. There are few voluntary organizations and no political parties which have such a broad base and such a wide coverage, although outside of the core membership will entail little beyond paying, grudgingly or otherwise, the subscription.

TABLE 6.2: Membership distribution of residents' groups

Membership	*n*	*%*
< 200	43	47
200–400	20	22
400–800	11	12
800–2,000	11	12
> 2,000	6	7
	91	100

The social base of the membership is shown in Table 6.3. These figures were based on asking respondents the predominant type of member. Ideally, full surveys of selected groups would provide a much better assessment, but failing this, our admittedly second-best solution is not entirely invalid since the core members interviewed are likely to have a good knowledge of the group membership. The statistics should therefore be seen as provisional indicators rather than precise figures. The bias shown in the table may therefore have some imprecision. However, given the findings of other studies (see Lowe and Goyder, 1983) the results are indicative of the selective nature of residents' group representation. This bias has been found in many voluntary organizations. Where the

TABLE 6.3: Predominant occupational status of membership

	Total sample (%)	% Stoppers	Stoppers -getters	Getters
Professional, white-collar	54 (59)	82	50	35
Skilled and semi-skilled	8 (9)	2	6	18
Unskilled	1 (1)	15	44	3
Mixed	29 (31)	1	–	44
	92 (100)	100	100	100

organization represents wide interests the selective nature of the membership may be incidental. We need to distinguish between the articulation and representation of interests. For example, because some voluntary group aims are articulated by middle-income members does not necessarily mean that low-income interests are not represented. However, in the case of residents' groups we are dealing with very parochial concerns and there is a close connection between articulation and representation. The selective membership of residents' groups is indicative of only certain interests being both articulated and represented.

There are differences between residents' groups. As Table 6.3 shows there is a marked difference in dominant status type between the three categories. The stoppers are overwhelmingly professional, the stoppers-getters only slightly less so while the getters are a more mixed bunch reflecting the importance of council and lower-income private estates in this category. The differences are further reflected in the dominant tenure type shown in Table 6.4. The stoppers are predominantly owner-occupiers while the proportion of other tenure categories increases in the stoppers-getters and getters.

In summary, then, the voice of the stoppers is the voice of the middle-class, middle-aged owner-occupiers seeking to protect their physical and social environments. This voice is also strong in the other two types of residents' groups but it is not the only one. If you listen you can hear the sound of younger owner-occupiers in new estates and inner-city areas and the demands from tenants' groups in council estates; here the concerns are with not only protecting but also enhancing the local physical and resource environments.

TABLE 6.4: Dominant tenure type of membership

Tenure	Central Berkshire (%)*	Residents' associations (%)			
		All	Stoppers	Stoppers-getters	Getters
Owner-occupation	65	68.4	88	67	47
Council housing	22	5.4	–	6	12
Private renting	8	–			
Mixed	–	26.2	12	27	41
Other**	5				
	100	100 (n = 92)	100 (n = 40)	100 (n = 18)	100 (n = 34)

*Figure refers to households in permanent buildings (source 1981 Census)
**Includes housing associations and housing linked to employment

Money

The work of residents' groups requires money. The subscription is a basic source of revenue and almost three-quarters of our sample charged some form of subscription ranging from 50p to £4 per annum. A group like CPVA which has 1,700 paying members can generate an annual income of £3,400 from subscriptions alone.

There are other sources of finance ranging from donations and grants to fundraising ventures such as coffee mornings and bring-and-buy sales. The more active groups spend most of their time in organizing such events. Two-thirds of our total sample had sources of finance other than from subscriptions. The newer groups of Fords Farm and Erleigh Road have not yet managed to obtain other sources of finance, while the more active community-type groups like Katesgrove and CPVA use all three sources. Dee Road is a typical council estate reliant on grants from the local authority and its own fundraising while the more sedate Ascot Society relies on the more leisurely source of donations; not for them it seems the round of jumble sales and sponsored walks.

Advice

Money is not the only resource. In dealing with the full-time professional staff of local planning authorities and builder-developers, residents' groups can profitably call upon the services of expert skills. The ability to obtain such advice is a major resource. In our interview we asked respondents if they used or had internal or informal (i.e. free) sources of advice from planners, solicitors, architects or others. Just over half of the total sample did not have or use any internal/informal source. When we broke it down by type of group we found that the stoppers used these sources more than the other two types. Consider our case studies: the Ascot Society can call upon the advice of members who are planners, solicitors, architects, county councillors and two MPs while the Hurst Society receives advice from planners. The Erleigh Road group have solicitors, architects and former planners on their committee. Dee Road in contrast has no informal access to such skills although it obtains free advice from a solicitor who advises a number of resident groups in central Reading. Katesgrove has no internal sources but it obtains free technical advice from members of Reading University. Fords Farm has, as yet, no internal sources of advice and, as befits the most organized group, CPVA gets internal advice from planners and architects as well as employing its own solicitor.

The stoppers groups, due to the nature of their membership, can generate advice internally. Among the middle-income groups there is a clear distinction between the long-established CPVA, able to draw upon local support, and the more recently established Fords Farm RA, which has yet to build up a network of local contacts. In the lower-income Dee Road RA there are few local skills directly related to land use planning.

External contacts

In order to achieve their aims residents' groups need to make contact with a variety of public and private agents. In this section we will consider their involvement with three sets of agents.

Links with other groups

Residents' groups are often portrayed as being in intense competition with each other as they compete for limited resources or try to palm off unwanted developments down the road to the neighbouring group. Although such instances do occur there is cooperation as well as competition between groups at both a national and local level. Of the total sample, only 29 per cent were in touch with national groups such as the Civic Trust, the National Federation of Community Organizations and the Council for the Protection of Rural England. Contact was usually initiated by the home group. About a quarter were formally affiliated to the national group, a quarter hoped to collaborate over a local issue, and the rest were looking for information and/or advice. By contrast, a far larger number (74 per cent) were in touch with other local groups. Reasons for seeking contact had a different emphasis from the reasons for contact with a national group: 64 per cent wanted to collaborate; 22 per cent wanted information and/or advice; 10 per cent wanted both of these and only 4 per cent were formally affiliated.

For our sample it seems that there are generally high levels of cooperation experienced between voluntary groups at both a national and a local level.

There were some differences between the groups. The national connections of the stoppers and some of the stoppers-getters centred on such groups as the Council for The Protection of Rural England especially for the rural stoppers and the Civic Trust for the more urban stoppers. For the getters and the remainder of the stoppers-getters the national groups looked to for advice were such bodies as the National Association of Community Organizations. At the local level all groups tended to link up when issues affected their mutual interest. There was standing arrangements for the getters, and even for the stoppers, where you would expect more competition, there were instances of cooperation. The Hurst group, for example, was a member of the Thames Valley Forum and the Berkshire Environment Forum (a loose grouping concerned with physical environments) and represented all the villages when it made comments on the Central Berkshire Structure Plan. At the local level, then, there are a whole series of overlapping concerns which are expressed in alliances and affiliations of varying degrees of permanence and success. Figure 6.8 shows the overlapping connections between a sample of groups interviewed; the impression given is of a dense circuit of local connections to be used in pursuit of common goals.

Contact with local and national politicians

Resident groups also maintain regular and long-standing contact with local politicians. Almost 98 per cent of groups had links with their district and parish councillors and in four out of every five cases the contact had been in existence for over five years. The reasons for such contact are clear. For the resident groups the elected members are a voice in the court of the local authority representing local opinion and in this role they play the part of place agent. For the councillors it is electorally advisable to pay attention to local articulated opinion. If you look back to Table 4.4 you will see that community groups were ranked second in degree of importance by planning councillors. After the officers, resident groups provide a major source of influence on the planning decisions made by local councillors.

The success of any external contact is facilitated if there is a personal link. In our interview we asked respondents if they had members on their group who also held posts in either parish, district or county councils. Almost a half (43 per cent) had such representation although the largest proportion (27 per cent) had only a parish council contact. The stoppers have the largest degree of representation. This finding, however, needs to be treated with some caution. The stoppers are predominantly rural and suburban-based groups and the parish system only exists in the suburban and rural areas. Moreover, councillors may find it expedient to be a member of a local residents' group without sharing its aims or goals. Just because a residents' group has a councillor as a member does not imply that it has better access to effective power than a group without a councillor. Contact with MPs was rarer. Only 57 per cent had got in touch with their local MP at any stage in the last five years. The contact invariably

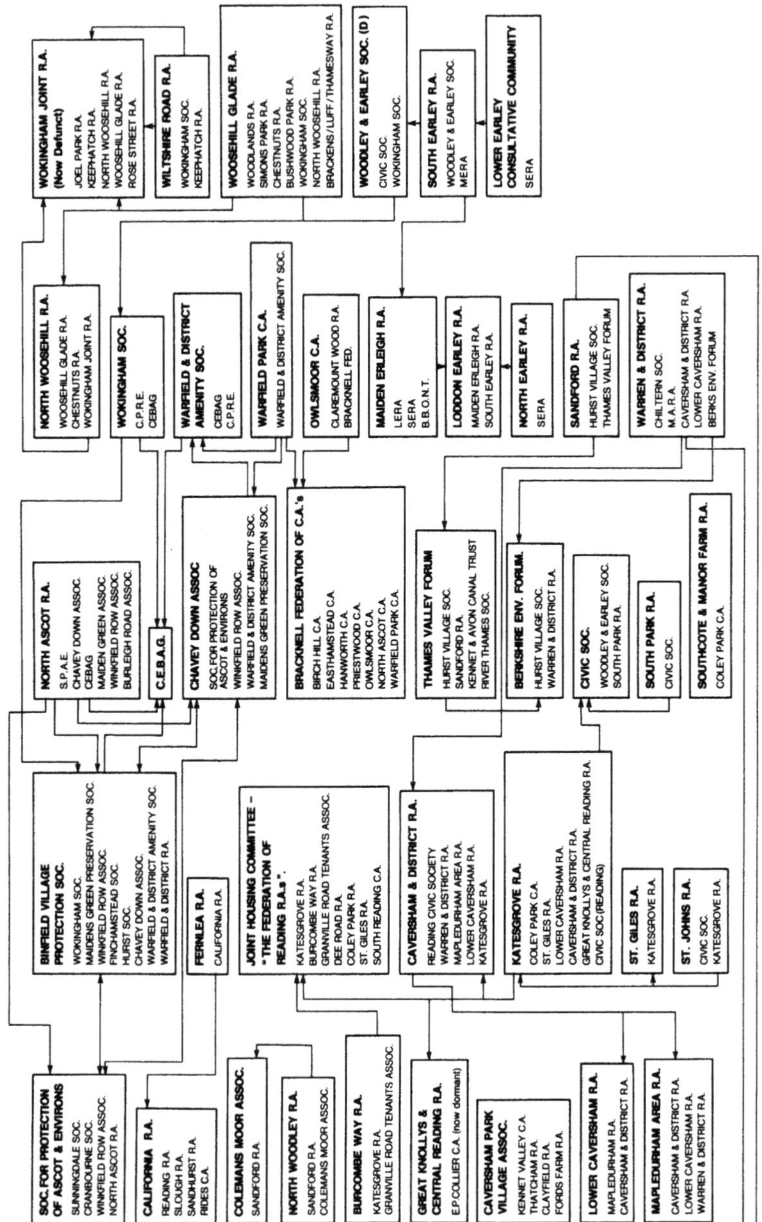

FIGURE 6.8 *Local linkages of selected residents' groups*

proved to be less than useful as they had little effective control over local planning issues, and over the issue of Heseltown the local MPs were ineffective. Between the power of local planning authorities and central government departments the ordinary backbench MP has little room to exercise any real power.

Contact with the planning system

The resident groups of Central Berkshire are involved with the local planning system in a number of ways. As Table 6.5 demonstrates, the most important form of involvement is the *monitoring of planning applications*. Such monitoring is an essential element for any group wishing to restrict or modify developments in their area. Not surprisingly over four-fifths of the stoppers are involved while just over a half of the getters are involved in examining all planning applications. Groups place objections to unwanted developments. However, it is difficult to measure the precise impact of residents' group objections. In only 14 per cent of cases were planning application objections by residents' groups upheld. But just because the outcome coincided with the group's objective does not necessarily mean that the group's action produced the outcome. It is difficult to measure the relative success of residents' group activity.

TABLE 6.5: Residents' groups and the planning system

Type of involvement	*% groups involved*
Planning application	72
Local plans	50
Development briefs	46
Public inquiries	42
Structure plans	41
Heseltown	33

Much of residents' group contact with the local planning system is through statutory procedures of public participation in *local plans* and *structure plans*. There is also the official channel of local authority–resident group consultative committee *comments on the development briefs* for the large new residential estates of Lower Earley and Woosehill. Here residents' group activity is formalized into making responses to official plans. *In public inquiries* residents' groups are brought into direct contact with developers and landowners. Whereas structure and local plans mediate development and containment pressure, public inquiries present the competing interests more clearly and directly. In our sample 42 per cent of groups were involved with public inquiries and in 50 per cent of cases the result eventually went the way of the resident groups. This higher 'success' vote is probably due to the fact that it is the contentious issues which reach public inquiries and therefore residents' groups desires may be more closely aligned to broader political and planning objectives than the case of objections to the more mundane planning applications. However, the concept of success implying residents' group influence independent of wider influences needs to be treated with caution.

Although Table 6.5 shows that only 33 per cent of groups were involved in Heseltown, this is a surprisingly high figure. The other categories refer to routine or recurring activities while Heseltown refers to a single issue. The main involvement came in April–May 1982 when the public consultation document *Central Berkshire – Land for 8,000 houses: The Choices* was published. Many residents' groups gave written replies to the proposals. Hurst Village Society, for example, wrote that it was strongly opposed to the identification of land for 8,000 houses. But of the various options, it favoured option 5 which put 2,000 houses at Spencers Wood, 1,800 houses at north Wokingham, and 2,000 houses at north-east Bracknell with the remainder spread through Burghfield, Binfield, Theale and Winnersh. Katesgrove Residents Association preferred option 2 which put 2,000 houses at Theale, 1,400 at Winnersh and 2,700 houses at north-east Bracknell because it minimized expenditure and had least impact on Reading town centre. Some residents' groups were particularly affected by some of the Heseltown options. In option 1 and option 5 3,300 and 2,000 houses were allocated to Spencers Wood, a small village several miles south of Reading. The local residents' group is the Spencers Wood Residents' Association (SWRA). This group, classified as a stopper in our classification, was formed in the late 1960s as a protectionist group seeking to monitor and propose alternative routes for the M4 motorway south of Reading. Interest and membership peaked in 1982, after a period of dormancy since 1974, in response to the Heseltown options.

A highly organized campaign was pursued by SWRA in their efforts to deflect this development. All residents in Spencers Wood (members and non-members) were given a copy of the public consultation document and the accompanying questionnaire to be returned to the county council. The committee urged that all members fill in the questionnaire and encourage neighbours to do so – they even suggested the way in which to fill in the questionnaire.

The second major line of attack was the production and publi-

Block 22 The fight against development

Mr Smith is a computer operations' manager. He is in his thirties, married with two children and lives in a house valued by himself at between £45,000 and £50,000. He has lived in this house for the past thirteen years. He is a member of a stopper group (a village protection society) as well as being a parish councillor, and a member of Bracknell District Association of Local Councils, Berkshire Association of Local Councils Executive and the Council for the Protection of Rural England.

He was a founder member of the village protection society established in 1975 after the Bracknell Feasibility Study of that year suggested a large concentration of housing in the village. The society was founded to oppose the implementation of this plan and oppose any other attempt to destroy 'our village community'. Mr Smith was secretary of the society from 1975 to 1978 and has been chairman since 1978. When asked what he got out of being a member, he replied, 'Knowledge of local government, planning etc, and plans for local neighbourhood, meeting village people and helping to advise others.'

The success of the society to date has been 'the general level of opposition to DOE's amendment to the Central Berkshire Structure Plan. . . . People in (the village) are more aware of the threats to their village.' Although he adds, 'If we had known in 1975 what we now know then the fight may have been more successful. The involvement of other committee members has been good but its continuation needs new blood.'

cation of a report-style document entitled *Spencers Wood: The Case Against Major Development* submitted to the County Planning Officer in 1982. Copies were also sent to every county and district councillor, every Shinfield parish councillor and other interested bodies (such as the CPRE). The document was extremely well laid out in a similar format to a planning document. It listed the issues – the principles of development, impact, cost, implications – and went on to discuss these matters considering options and identifying those favoured by SWRA. In addition SWRA committee members monitored meetings of the Environment Committee and full council of the County Council and sought to gain further support from other local groups (such as the CPRE, Wokingham Society, parish councils).

The SWRA case against options 1 and 5 were that

(1) existing services were inadequate and expansion of services would be costly. Developers would be unlikely to finance the high cost of public works;
(2) existing highways were already overloaded;

(3) the extra population would swamp existing communities and the town and country divisions would be submerged.

Of the five options, SWRA preferred option 4, which concentrated development in the Reading/Wokingham/Bracknell corridor.

It is interesting to compare the option choices of SWRA with that of Hurst Village Society and Katesgrove. Each and every group was opposed to the 8,000 houses in general but when faced with the need to make a choice sought to deflect the development to other areas. When the *no-growth* response fails, groups resort to the *no-growth-here* stance. Ultimately, of course, all the residents' group activity seemed to have little palpable effect. The final decision taken in late 1982/early 1983 by the Environment Committee of Berkshire County Council which went for a north-east Bracknell option was due more to party political and planning gain considerations than to residents' group activity. However, the activity of residents' groups structured the decision making environment of council members.

Contacts during campaigns

A very clear pattern of contacts emerged; most groups first contacted the district council, then the county council and occasionally the parish council, with most petitions, objections to planning applications, and requests for routine maintenance and community facilities going to the districts. A small minority contacted their MP or a central government department and a few groups collaborated with another local group during a campaign.

The councillors, builder-developers and other parties were contacted by the residents' associations largely within a formal framework. All those involved in wrangles over maintenance, community facilities, builder-developers' plans, and 91 per cent of those compiling petitions, used letters and meetings with the other parties to put their point across. Only eight groups held demonstrations to further their aims, and none of the groups took part in any illegal action. Most of the groups in our sample seem to have been embraced by the established institutional framework. This may reflect either their incorporation into the existing framework or their relative success in getting their aims realized, or a mixture of both. The lack of militant action can reflect both incorporation and relative success.

Block 23 Community group activities and linkages

Mrs Jones is a housewife in her thirties. She is married with two children, aged 10 and 7. She has lived in her present house for the past eight years. Her house is valued at between £50,000 and £70,000. She is a member of a getter group. In an average week she spends about nine hours on residents' group affairs.

'What were your reasons for joining the group?'

'I wanted to learn more about what was happening locally.'

'What is the nature of your involvement?'

'I was approached and asked to be a street rep by the membership secretary. At a subsequent AGM I was asked to stand by a friend who was retiring from the committee. I served one year as an ordinary committee member followed by a year as membership secretary and am now in my second year as chairman. During the last five years I have been also, and still am, a street rep and area rep. (Reps deliver the newsletters and collect subscriptions and area reps organize the street reps. Each area rep has about ten street reps.)'

'What do you get out of belonging to the group?'

'An insight into the workings of local government. The RA has been going since the estate was first developed (about sixteen years) and most of the new estate problems have ceased although the building within the area still continues.

'I stayed on to help the committee, otherwise all new people, come to grips with the problems two years ago. Two years in the chair is long enough.'

'Internal affairs?'

'We have a very efficient network of street reps and area reps, some sixty odd people which I remodelled when membership secretary. Previously we had very poor records of membership and rather sloppy bookkeeping. This is now very well organized. At any one time the membership secretary can quote the names of all members and say when they joined and give their membership number and full address.'

'Contact with other groups?'

'We have had very little contact with other groups in recent years. We disbanded the Joint Residents' Association last July because one person was writing to the press expressing his own views but calling himself chairman of the Joint Residents' Association. Eight member groups came together to stop him by disbanding the groups. We agreed to hold informal meetings about every three months. The first was well attended but for the second only one person turned up!

'As regards council matters, by and large our views have been taken favourably into account and we do not feel we have "lost" an issue on which we have made a stand.'

During campaigns, the majority (76 per cent) of groups are in touch with the media. The predominant link is with the local newspaper while a few have presented their case on local radio and television. The importance of publicity is recognized by the residents' groups; it gives their campaign a higher public profile and puts extra pressure on the other actors in any issue. The most successful groups are the stoppers who have managed to combine their own local concerns with wider public interest through the medium of general environmental concern.

Consequences

While it is relatively easy to note the aims, internal structure and external contacts of residents' groups, it is very difficult to ascertain their success in achieving their aims. Between the desire of the group and the final outcome lie a myriad of other interests, conflict and bargaining. Some of the range of mediating factors are highlighted in the case studies presented in Chapter 7. It is difficult, therefore, to reach conclusions about the precise impact of residents' group activity. While a precise balance sheet cannot be drawn up, it is possible however to make three general points.

First, the stoppers have been influential in creating an articulated and powerful no-growth lobby in Central Berkshire which has sensitized local members to the issues of growth containment and growth deflection. The actions of the stoppers places growth minimization higher than growth generation on the agenda of planning in Central Berkshire. Given the ultimate control of central government in overriding local interests, as the case of Heseltown clearly shows, this means that elected members have to implement growth strategies and pro-development planning decisions in a climate of anti-growth. The members have responded in two main ways:

(a) they have sought to deflect it elsewhere. Development often takes the line of least resistance;
(b) where this is not possible they have attempted to lever something out of development. There are many reasons for the growth of planning gain agreements but an important one has been the need for members to show positive benefits in development to articulated anti-development opinion.

Second, the getters are becoming important in two respects. On the one hand they may become the stoppers of tomorrow putting extra pressure on the planning system to restrict and deflect growth. On the other hand, given their capacity for community self-help, they will be least affected by reductions in public services. The impact of the decline in public services will be differentially felt with the more affluent getters least affected. Given the drift of public policy and government expenditure, sources of inequality will become apparent between those communities able to generate self-help and those unable to do so. This distinction will not be simply one of income and status, but of income, gender, status, length of residence and all those other factors which create a community feeling. With their success in mobilizing internal resources, the getters are best placed to counter some of the effects of reductions in public services.

Block 24 'You win some, . . .'

In this research project we were very reliant on the goodwill of a large number of people to give up their spare time to answer our questions. After the survey of residents' associations we called on certain respondents in a follow-up survey. The respondent whose letter we reprint and who was a key member of a stopper group had obviously had enough.

Dear Sir,

I am sorry, but the interview I gave you is the best I can manage, and that I did reluctantly, for I sincerely doubt the value of the research you are doing and would be most surprised if you reach any conclusions different from those that common sense would dictate.

A decent answer to your questions would take ages and in many respects would be very subjective. Very few local groups such as that of which I am chairman have issues to justify Qs 4 and 5 [concerned with internal affairs and links with other groups].

In general it is fair to say that pressure groups develop because some people are concerned enough (usually for a mixture of personal and community reasons) to try and do something about a particular issue. Most of the people who are active tend to be middle-class because understanding the issues demands a certain level of understanding and *broadly* speaking general ability leads to a decent education and a reasonable standard of living. (I am thinking particularly of subjects such as pressure on the countryside and on small communities, which largely concern me.)

Sorry not to be more helpful. Who on earth is *paying* for the research? Not, I hope, the taxpayer!!

Finally, we can note that residents' groups represent specific interests. The concerns which they express are those of many owner-occupiers and some public sector tenants, either seeking to restrict further growth or to generate community facilities. They do not fully articulate the interests of the homeless or the unemployed, while the voice of private sector tenants is scarcely heard. Not all issues which affect local communities are thus placed on the agenda of public discussion and political decision making by residents' groups.

7 Stories and tensions

'A place has almost the shyness of a person, with strangers; and its secret is not to be surprised by a too direct interrogation.'

(A. Symons, 1903)

In the previous chapters we have considered the different agents (the housebuilders, planning authorities, parishes and resident groups) individually, examining their actions in relative isolation. The reality, however, is of continual reaction as the different agents interact with one another. The problem is how to convey this agent interaction in an accessible form. In this, the final chapter, we will seek two solutions. On the one hand we will present three case studies which illuminate various aspects of the real-time negotiation, compromise and conflict generated by proposed residential developments. We have tried to make these case studies as comprehensive as possible by interviewing as many of the different actors in the separate dramas as we could. On the other hand we have also sought to make some generalities by reconsidering some of the tensions identified at the end of Chapter 2 but now, after four chapters of detailed material, on the basis of a much firmer grasp of the operation of, and constraints on, the different agents.

A parish initiative

> 'I accept a planning process in which the individual citizen, whether alone or acting as a pressure group, has a fundamental right to influence the decisions taken in the area in which he lives.' Michael Heseltine in a speech at York in 1979, quoted in the *Wokingham Times*, 6 December 1979

In November 1982, the parishes of Binfield and Warfield in the district of Bracknell jointly published *Goodbye Rural Berkshire* – a document which challenged the DOE's method of calculating housing land supply and its policies towards national planning and the South East. In the following month, December 1982, the parishes of Binfield and especially Warfield were designated by Berkshire County Council to be the recipients of one-half of the Heseltown 8,000 houses allocation. Their document was influential if a little late. The lead-up to the publishing of this document is useful in illuminating the role of at least two parish councils in Central Berkshire and their relations to their wider community.

Increasing pressure

The spectre of growth in Central Berkshire is not a new image to many of its inhabitants. A number of regional and county reviews since the late 1960s had clearly indicated the possibility, or even inevitability, of long-term growth in the area. This included the designation of Area 8 (Reading, Wokingham, Aldershot and Basingstoke) as a major growth area in the 1970 Strategic Plan for the South East.

In the early 1970s, there was relatively little community action in planning affairs. Such action was however encouraged by the Skeffington Report of 1969 and in the mid- to late 1970s, structure planning activity stimulated many groups. Two such groups were the Warfield and District Amenity Society formed in 1974, and the Binfield Village Protection Society formed in 1975 in response to the structure plan identification of land for development at Amen Corner an area of about forty-five hectares just over a mile west of Bracknell. Other groups formed but generally their actions were isolated and uncoordinated.

However, in October 1979 when Michael Heseltine, fresh to the post of Secretary of State for the Environment, announced his intention to amend the Central Berkshire Structure Plan to allocate land for a further 10,000 houses, there was a concerted drive from a number of voluntary groups. Ten village and local amenity societies (some of which were resurrected for the issue) from the threatened areas of Bracknell and Wokingham districts combined to form the Central Berkshire Action Group (CBAG) in November 1979. In the six-week consultation period for the proposed amendment, there was feverish activity. A local paper with a history of attachment to the idea of protecting 'the green fields of Berkshire' instigated a petition to oppose the amendment. Through its membership, CBAG had ready access to the media generally (at least one society member was a television producer) and the organization obtained slots on both BBC and Thames TV news programmes. The campaign against the amendment culminated on the final day of the six-week consultation period when CBAG hired Westminster Hall for a press conference and presented the petition, complete with model of Berkshire, to prime minister Margaret Thatcher in Downing Street. Thus from an early stage, the campaigners had sought to give the case of Central Berkshire an airing nationally. They considered that the issue at stake was large and best fought against the UK context of the North-South divide. But it would be naive to assume that self-interest was not the real issue.

In April 1980, Heseltine formally issued the amendment to the Central Berkshire Structure Plan. Perhaps as a result of the opposition articulated in Central Berkshire, the original amendment for land for an additional 10,000 houses was reduced to 8,000. The concession was not warmly greeted. The editor of the local paper which had generated much of the opposition openly stated in an editorial that the fight had been lost. Welcome for the amendment came from few quarters – it was ironic that most opposition to the Conservative government's decision came from the local Conservatives and that it was a Labour councillor of Bracknell District who volunteered; 'It's a good idea and I hope a lot of the extra houses come to Bracknell.' The Labour councillor hoped that Bracknell's housing waiting list would be shortened by development in areas such as south west Binfield and Amen Corner.

Pressure resisted

Others, however, did not consider the fight against 'Heseltown' to be lost. The driving force behind much of the CBAG operations was a member of the Binfield Village Protection Society. In 1980 he stood as an Independent for a Binfield Parish Council by-election and was elected with a large share of the vote. Another colleague was also elected and together they went on to the planning committee of the parish council. Over the following year they made a point of writing frequently to Berkshire County Council and its Environment Committee stating their case.

In 1980, the CBAG activist also joined the local branch of the CPRE and soon came to be on its executive committee. Thus in five years he had come to occupy important positions in four local organizations from which the case against Central Berkshire as a growth area could be pursued.

Most of the activists of north of Bracknell now claim that even though the campaign was ostensibly against 'Heseltown' they recognized from an early stage that it would be unsuccessful. None the less, they considered it a valuable exercise to indicate to central government and developers that Central Berkshire was an unwilling recipient of housing development. By demonstrating their resistance, they hoped to provide the context for future development.

In the spring of 1982, Berkshire County Council, against the advice of the DOE, went to public consultation to help them to identify the sites for 'Heseltown'. From an early stage, the DOE raised objections fearing that the consultation exercise might be interpreted as a consultation over the need and not the location of the 8,000 houses. None the less, the county forged on.

During the consultation process, however, it came to light that the case against the principle of the 8,000 houses might still be fought. There are conflicting views as to how this came to light but the most consistent view is that a Council for the Protection of Rural England (CPRE) source revealed to parish councillors of Binfield and Warfield that there was a statutory procedure by which the county could appeal against Heseltine's amendment. Personal contacts had by now been established between leading parish councillors of Binfield and neighbouring Warfield. The possibility of appeal against the amendment was publicly aired for the first time by the chairman of Warfield Parish Council in a local radio discussion with the Chief Planning Officer of Berkshire, the county spokesperson of the Conservative and Labour parties and the local MP.

The revelation of the possibility of the appeal caused a stir. The leader of the main political party (the Conservatives) of the hung county council was surprised, and concerned that his CPO had not informed him of the possibility. He immediately instructed his CPO to embark upon the procedure. It was eventually decided that the council should press not for the complete abandonment of the amendment, but rather a reduction of its allocation from 8,000 to 4,000 houses.

Meanwhile, the consultations over the location of the sites of Heseltown continued. The individual responses of Warfield and Binfield parish councils were to suggest not one of the five options indicated. Rather, they proposed a Zero Option – that none of the 8,000 houses should be located in Central Berkshire. To further propagate their views, two of the key parish councillors of Binfield and Warfield forwarded themselves for important posts on the Berkshire Association of Local Councils (BALC). They were elected and were instrumental in June 1982 in calling a BALC meeting at Shire Hall (the county offices) at which BALC deplored 'the total failure of the Secretary of State for the Environment to take proper account of the numerous representations being made to him rejecting his belated and divisive edict that an additional 8,000 houses be provided in Central Berkshire'.

A report is produced

However, they recognized that to be effective the fight needed to be carried to an even broader front. They jointly agreed that it might be useful to engage a planning consultant to research the evidence and produce a substantive case against the 8,000 houses. After consulting their CPRE contact who had unearthed the amendment appeal procedure, they were put in touch with one of the contact's ex-college contemporaries. After an informal interview, James Mackie was employed for six to eight weeks to act as a freelance planning consultant to draw up and present the case. The costs were to be borne equally by Binfield and Warfield parish councils and each of the Binfield and Warfield Village Amenity Societies. The eventual bill (including publication costs) totalled around £2,000.

Mackie duly produced a draft document. It was favourably received by the parish councils although they did have some reservations. They felt that the draft was too academic and that it needed restructuring and honing (one of the councillors was a professional report writer and so skills were readily available). While the Binfield councillors felt that the conclusion should be stated at the outset to ensure that the readers would not miss the thrust of the argument if they failed to finish reading the booklet, the Warfield councillors were concerned that by stating conclusions first many readers would be tempted not to read on. The Binfield view prevailed. The councillors also wanted the examples made less complex and the mathematical content of the text reduced. Interestingly, too, they felt that the tone of the draft was at times too tough. While they felt that 'middle-class ghettos' and 'blackmail' might be accurate terms, they thought that they might backfire and alienate those whom they wished to impress. Two of the Binfield members were largely responsible for the rewriting of the draft but the original author was clearly displeased by their editing and felt that his work had been diluted.

The report (Mackie, 1982) was eventually published in November 1982. Although it was 'a considered response to . . . structure plan housing land allocations in Central Berkshire' (p. 7) and was submitted to the County Planning Department, the report was really aimed at the inconsistencies and policies of the DOE. The report pursued two main themes: first, that the DOE's method of calculating land supply (Circular 9/80) is inconsistent and biased in favour of developers and that in any case, Central Berkshire had at the end of 1982 at least a 6.5-year land supply (well above the 5-year minimum); and second, that the promotion of Central Berkshire as a growth area is outdated, detrimental to

Inner London and wasteful of the capital's infrastructure, and detrimental to Central Berkshire because it was wasteful of its agricultural land and of the finance required to provide its infrastructure. In the event, the thirty-page, illustrated booklet report was very readable: the report was fronted by a two-page summary, ended by four pages of notes and references, with the meat of the argument sandwiched between.

To publicize the report, the parish councils of Binfield and Warfield compiled a mailing list to cover a wide range of interests: local district and county councillors, Berkshire parish councils, local MPs, the chief planning officers of counties in south east England, the media, and various statutory bodies and individuals. The report was sent free of charge to these interests and at local planning meetings in Berkshire copies were often freely available.

From a print run of 2,000 it is thought that 1,000 were distributed. The total cost of £2,000 was met by the two parish councils (through their parish precept) and by their respective village protection societies. Along with the publication, a note was sent to parish councils indicating that donations would be welcomed. Contributions ranged from £5 to £100 and eventually some £300 was received.

The response to the document was mostly favourable, if a little subdued. However, it was clear that the county planning officers were somewhat aggrieved. They have claimed that in part the report is inaccurate and the CPO did in fact write to Mackie complaining that the help which his officers had given had not been credited. Mackie's response was that in fact the officers had not been very helpful.

The feeling amongst the parish councillors most closely involved in the production of the report is that they have not endeared themselves to those in power – at the county and the DOE. They intend to continue the battle even though they doubt that their actions have yet had any tangible benefits. To fight central government over the issue, they feel that a dramatic and radical Clay Cross-style action would be necessary – but they consider such a stance unlikely. Their view is that the officers at the county are very powerful in relation to their committees which have found it difficult to cope with the issues and technicalities of planning. They feel that the members have accepted the situation as presented by the DOE too readily.

The report is ignored

A powerful irony concluded this particular parish council initiative. On 14 December 1982, one month after the publication of *Goodbye Rural Berkshire*, the full County Environment Committee convened to debate the proposals for the allocation of the land for the 8,000 houses. The public consultation exercise had itemized five possible options for the spread of development. None of these were recommended to the committee. Instead one of another set of options requested by an environment subcommittee was recommended. That particular option, Option 9a, which was instigated mainly by the Labour party environment spokesman allocated to NE Bracknell (bordering Binfield and Warfield) no less than one-half of the 8,000 houses allocation. These 4,000 houses were 1,300 more than those suggested for NE Bracknell by any of the original five options. In an attempt to soften the blow, the development at NE Bracknell was to be phased in last (from the early 1990s). At a meeting of the full county council in January 1983, support for 9a was confirmed.

Postscript

For a while, Bracknell District Council hoped (mainly in public) that if the appeal against the structure plan amendment was successful, North East Bracknell would be the area to be reprieved. But by the end of 1983 this hope had faded, following contacts with the Secretary of State and his senior officials at the DOE. The parish councils continued their struggle however, despite the apparent failure of their initiative. During late 1983 and early 1984 they met with the various residents' groups and amenity societies in the area to consider their situation and future tactics. In addition to the 4,000 houses scheduled for the early 1990s, concern was growing over the forthcoming review of the Central Berkshire Structure Plan, which could herald yet more growth for North Bracknell.

The northern parishes discovered a powerful ally for their fight in the newly elected (May 1983) MP for Bracknell. He began to campaign for the removal of the growth area status from Bracknell (and, indeed, from Berkshire as a whole), with far greater vigour than that of the other local MPs. This gave great encouragement to the parish councils as they prepared for the next phase of the anti-Heseltown fight. In January 1985 the Northern Parishes Action Group (NORPAG), comprising most of the parish councils and local groups around the northern fringes of Bracknell, was formed to oppose both the 4,000 houses and any additional development.

This body provided united opposition to the drafts of both the structure plan review and the North Bracknell Local Plan, which were released for public consultation in early 1985. Some Bracknell councillors condemned the NORPAG objectives as irresponsible and unrealistic, but the pressure against development mounted. With the prospect of future Bracknell allocations in the reviewed structure plan, the district council began to see advantages in public resistance to the 4,000 houses. Therefore on 29 March 1985 the council formally resolved to reject the NE Bracknell 'Heseltown' allocation and also called upon the government to halt development, even though their own local plan for the area was in progress. The preparation of *Goodbye Rural Berkshire* had proved to be only one phase of the ongoing struggle to protect Binfield and Warfield from development.

Burghfield Common

This case study concerns two sites of land totalling about sixty-eight acres, which lie on the opposite sides of the main road through the village of Burghfield Common, five miles south of Reading. Pressure from private developers for the release of these sites spanned a ten-year period before planning permission was granted. The issue of residential development involved most of the major actors within the planning process: Berkshire County Council; the district council; the parish council; the Secretary of State for the Environment; housebuilders of different sizes; and a firm of planning consultants. The study concentrates on the final twelve months of the ten years, when the conflicts between the various actors were brought to a head, and the sites were released for development.

Background

The history of growth in Central Berkshire has been outlined elsewhere. It is sufficient for this example to note that an optimistic assessment of future economic and population growth formed the basis for the Area 8 strategy, which followed the designation of the Reading–Aldershot–Basingstoke subregion for major expansion in the 1970 Strategic Plan for the South East.

One zone considered for housing growth within the Area 8 strategy was the area to the south and south west of Reading, including the village under consideration. Although this proposal was soon dropped in the light of economic slump and opposition from the local authorities to continued expansion, attention had been focused on the settlements within the area. In 1973, one of the major national housebuilders submitted an application for an entire 173-acre farm on the south side of the road, only one-fifth of which formed the eventual southern site. Berkshire County

Council refused the application (the county being the only planning authority before the 1974 reorganization), but, after a public inquiry, the appeal inspector recommended that permission be granted. The Secretary of State for the Environment adopted a different position; whilst accepting the suitability of the site for residential use, he considered that the development of some 150 acres of housing would be out of scale with the village.

The first application for the northern site was submitted in late 1975 and refused by the district council on the grounds of insufficient demand for new land release. An application for the southern site by a locally based regional developer quickly followed. This comprised only the area indicated as more acceptable in the 1973 appeal decision by the Secretary of State. As soon as the statutory period for determining applications had elapsed, the developer, advised by a firm of planning consultants, appealed on the grounds of non-determination (a deemed refusal). The appeal decisions for both sites were made in 1978, and although the refusals were upheld, the terms of the appeal decisions represented an important change in the planning state of the land.

The suitability of the sites for residential use was confirmed, and the inspector considered that they should be released once the need for land could not adequately be met from other commitments. The appeals were therefore dismissed on policy grounds alone. The arguments of local residents, the parish council and district councillors against future applications on the grounds of loss of rural character, detrimental impact upon the village and inadequacy of services and infrastructure would be of no consequence. The defence of the local authority that the sites were rural and should remain undeveloped had been breached, and the principle of development was therefore effectively lost. The only criterion to be considered at any future appeal would be the extent of housing land availability within the district, and the need for further release.

The 1982 applications

A further application for the northern site was received in late 1979 and refused early in 1980. The public appeal inquiry was eventually fixed by the DOE for September 1982. In April 1982, however, the planning context of the site was significantly altered by the publication of the document *Land for 8,000 houses: the Choices* by Berkshire County Council. The planning background of the village made it an obvious candidate for some proportion of the extra dwellings. Of the five options produced for public consultation, four of them included a 300-house quota for the village, with the area for possible development including both sites.

In the wake of this document two fresh applications were submitted. The northern applicant was a civil engineering company, who would sell the site to housebuilders once outline permission and the infrastructure agreements had been finalized. Although the local objectors regarded both developers as a common foe, the two companies were in competition to a considerable extent. It was by no means certain that a decision to release land within the village would free both of the sites for development. As in 1975/6, the northern application was received first, followed after a few weeks by the southern one. When a northern site application was submitted which appeared to have a chance of success at appeal, the southern developer followed suit in order to put pressure on the local authority and to shadow the rival company in the queue for appeal inquiries.

The timing of the applications before the county had decided upon the Heseltown locations served two purposes for the developers. They viewed the whole public consultation exercise as a delaying action by the county council, an attempt to postpone the release of land in the hope of a change of attitude by the DOE or a change of minister through a Cabinet reshuffle. The housebuilders also suspected that a formal attempt would be made to challenge the 1980 structure plan amendment. The appeal decisions which would follow district refusal would indicate the DOE's attitude towards the consultation exercise; either supporting the 'delaying tactics' or pre-empting the county and district choices. During the consultation period, a number of appeal decisions throughout Central Berkshire were regarded as tests of the DOE position. All the developers were aware that once the 'prematurity' basis for refusal was overridden on one site, most subsequent appeals would also be permitted until sufficient land had been released. It was therefore important to have appeals 'pending' in order to pre-empt the rush of applications which would follow any move by the DOE to bypass the county's allocation process.

The second reason for submitting an application before the locations were determined concerned the negotiating tactics over the individual sites. After a refusal (as seemed inevitable) by the district, the forthcoming appeal inquiry would act as a bargaining counter in discussions over a further application. If the village received a Heseltown quota, the district would then be under pressure to grant permission rather than face certain defeat with the imminent appeal.

The district response

The parish council, who had objected strongly to all the previous applications, and many local residents were opposed to both the 1982 proposals. So were the district planning councillors, although some of the senior members were aware of the implications of the previous appeal decisions and of the Circular 22/80 land supply requirements. The planning officers had much less of a personal commitment to the protection of the sites, but were still reluctant to release the southern one. This was not firmly contained within the structure of the settlement, and would constitute large-scale rounding-off. There was also the possibility of further applications on other parts of the farm if permission were to be granted for part of the holding.

There were however sound and pressing reasons for release of one of the two sites. The stock of land within the district having planning consent for housing had dwindled from a position of glut after the releases of the early 1970s, to one approaching the critical five-year mark. Once this supply of land had dropped below the five-year level, only very strong planning reasons could prevent a site being released, and the officers knew that these sites had no hope of success at an appeal. They realized that the planning consultants were aware of this precarious situation, even if not the precise position. In addition, a refusal on both sites might undermine the appeal position of any other applications refused by the district. The release of a 250- to 300-dwelling site would satisfy the five-year supply needs for a further period, and would safeguard other vulnerable and perhaps more important sites.

In view of this situation the officers sought to obtain planning permission for the northern site on the best possible terms. Discussions were held with the developer seeking the withdrawal of the appeal (against the 1980 refusal) due to be heard in September 1982 (only a few months away). Both the planners and developer recognized that the inquiry would be irrelevant, coming only a short while before the county council's allocation of the 8,000 houses. The decision would be a refusal based simply on prematurity grounds, but the DOE had refused to defer the inquiry on the grounds that it would be a misuse of the planning process. The developer agreed to withdraw the appeal and the officers indicated their intention to seek committee consent for the northern site.

Before the application appeared at committee, the officers attempted to convince selected 'policy upholder' councillors of the need to release land and of the inevitability of the site's development within a few years. The written report carried a recommendation of consent and a strong case was presented by the officers at the committee meeting. Despite their efforts, the previous appeal decisions and the precarious land supply position, the northern application was blocked by the committee. Although the policy upholders had some sympathy with the officers' views, most members were totally opposed to both applications. They had no intention of granting permission for a potential Heseltown site before the county had determined the locations for the 8,000 houses. This would have provoked heavy criticism from the local area. To compound the problem, if permission was granted before 1983 the site would not count towards the district share of the allocation and other sites might be required. The members were opposed to the contention of the officers that the time required to complete legal agreements would prevent a pre-1983 consent. Granting planning permission would place full responsibility for unwanted, and locally opposed, development on the district council. If however the county eventually allocated 300 dwellings to the village, and/or the DOE allowed the appeal, then the blame could be directed elsewhere. These considerations were of more immediate concern to the councillors than the quasi-legal appeal and five-year supply factors outlined by the officers.

The Heseltown allocation

During 1982, the five options for the additional 8,000 houses were presented for public consultation, and the responses analysed and discussed within the county council. In view of the district officers' positive recommendation for the northern site, the planning consultants realized that a 300-house quota from the county would exclude their southern site. The justification given by the county for their 300-house limit was for infrastructural reasons; only very costly works could overcome the water supply and sewage constraints. In order to increase the quota for the village to include their southern site, the planning consultants needed to remove this drainage constraint.

In common with a number of the private sector planners and agents within the area, the consultant planner handling the case was a former local authority planning officer. However, in this case the consultant was particularly well informed about the site and area in question, having conducted the Appeal case against the (same) southern site for the county council in 1978, before leaving the public sector. This had involved close cooperation with the senior district officers, and the consultant therefore had detailed knowledge of the site, and the senior personnel of both the county and district planning departments.

The planning consultants decided that the approach adopted by the county council towards the drainage constraint was not sustainable. They presented a case to the county arguing that no physical constraint to development actually existed, and that the quota could reasonably be increased to include the southern site. The water authority confirmed the contention of the consultants that with an additional pumping station installed at the developer's expense, then a statutory obligation existed upon them to service the site. The county planning officers conceded the case and revised their potential allocation figures. The district officers had been kept informed of these proceedings by the consultants, but had not been contacted directly. In discussion with their county colleagues as part of the overall 8,000 houses allocation, they expressed no opposition to the increase. The principle of development on both sides had been effectively conceded, both sites would have been released within a few years even without Heseltown, and a certain proportion of the 8,000 was in fact essential for the district's five-year supply situation.

The final allocation was announced by the county council in late November 1982, following discussions between the party leaders of the three political groups on the council. The final option bore only limited resemblance to any of the original five options, and included a 500 rather than a 300 allocation for the village.

The 1983 applications

Fresh applications for the two sites quickly followed the announcement of the final option, with the northern application preceding the southern by just over a month. The consultants timed their submission so that the other application would be considered first, but not actually determined, before their own application was received. The applications would therefore be viewed jointly as the Heseltown quota by the councillors, but the northern site would attract most of the opposition. Since it was the more acceptable of the two, it would be permitted, thus extending the built-up area of the settlement and establishing the southern site more clearly within the structure of the village.

The 500-dwelling quota appeared to clear the way for speedy consents, but two complicating factors had arisen. First, the date before which no major releases were to be permitted had been deferred by the county, after positive signals from the DOE, from 'pre-1983' to 'end-1983'. This presented committee members with a legitimate excuse to refuse, on prematurity grounds yet again. Second, it was clear by the early months of 1983 that a general election was forthcoming, in addition to the local elections in May. Political considerations were therefore likely to become more important as thoughts turned towards the ballot box.

In view of these two factors, the district officers found themselves in an awkward position. They saw the applications as clear-cut consents in the context of the existing planning framework, and were certain that the already outstanding appeals would be lost, together with the opportunity for local control and probably the costs of the inquiries. But despite all the factors pointing to consent they feared refusal decisions by the committee, taken either on principle or because it was politically expedient to postpone the evil day until after the elections. At a personal level, refusals would present the officers with hopeless appeals to fight, and in planning terms they would endanger other more important sites. Lobbying was therefore again carried out by the senior officers in the direction of two influential members of the planning committee: the chairperson and the ward member. All the planning aspects of the case were presented to them, together with assurances that greater control and more satisfactory details would be obtained from a negotiated permission than from a refusal and a lost appeal. The danger of having less than a five years' supply of available residential land was particularly stressed.

The two councillors, both influential within the controlling Conservative group on the council, were faced with a difficult political decision. The granting of planning permission for 500 houses could prove to be a dangerous handicap for the ward member, with elections only a few months away. The fact that all three authorities involved, central government, the county and the district, were Conservative-controlled was an electoral embarrassment, and a reluctant planning consent would only highlight the impotence of the district council in relation to the higher authorities. Since the developments had been staved off for ten years it was tempting to lead a committee refusal once more, postponing the final appeal decisions until after the elections, and shifting the blame for the actual consent on to the DOE. The ward member was also confronted by the opposition within the area. As a dual district and parish councillor there was pressure to vote in defence of the ward and in line with the parish council's strong opposition. On a controversial application the previous year the ward member had led a proposal for consent against strong local objections, and had been bitterly criticized in local press articles as a result. Both members however decided that the correct course of action as planning councillors was to accept the situation, consider the wider interests of the district and ensure that permission was granted on the most favourable terms possible. The inevitable had to be faced and both were of the view that their planning duty took precedence over political and personal considerations. A planning gain approach was therefore adopted.

The local opposition

The parish council was the principal channel of opposition to the developments. It had fought against all the previous applications, together with a number of public sector residential developments on the fringe of the settlement. Despite the formal allocation of the 500 dwellings by the county, the parish council mounted another campaign against the proposals. The local papers carried articles concentrating on the opposition to the development and the detrimental impact it would have upon the village.

Two themes were stressed in this opposition: the inadequacy of existing services, and the environmental impact. At a parish council meeting a seven-point letter of objection to the northern site was drawn up and sent to the planning authority. The extra 250 to 300 houses would, it was claimed, 'decimate an already overhoused piece of our rural heritage', congest the local roads and place unbearable strain on overstretched services such as shops, schools, electricity supplies and the drainage network. A prospective parliamentary candidate, who had supported a residents' action group on a previous planning controversy, joined forces with the parish council. A tour of the village was made with the parish chairperson, and included a meeting with the primary school headteacher where fears of overcrowding were voiced. A statement to local press reporters concluded that the plans for development showed little thought for the present, let alone the future needs of the community. However, despite the parish council campaign and the publicity given to the northern site, only ten letters of objection were sent to the district council by the time of the committee meeting. The local population appeared to have accepted the development as an inevitability following the Heseltown allocation, and the succession of applications since the early 1970s had dampened some of the enthusiasm to formally object to the local authority.

The first meeting

The northern site was the first to be considered by the committee. The objections of the parish council were reported to the councillors, followed by the case of the Planning Department. In a forcefully argued report the Senior Development Control Officer recommended consent on the basis of structure plan policy H4 (phased provision after 1983 for 8,000 additional houses), the recently approved county council locations (including 500 for the village), the needs of Circular 9/80 for a five-year supply of available housing land, and the suitability of the site for development as indicated by the 1978 appeal decision. Legal agreements were proposed to cover open space provision and maintenance, the provision of footpaths along two of the site boundaries, and financial contributions by the developer for access arrangements and improvements along the main road.

The ward member started the committee discussion, which subsequently involved all those members who would normally contribute at committee meetings. (Even on important applications council-filler members rarely enter committee debate.) Most of the speeches comprised three elements. The Department of the Environment was denounced for insisting upon the additional 8,000 houses and for its willingness to override the opposition of democratically elected local councils via the appeal mechanism. Pessimistic assessments were made about the physical and social effects upon the village of the additional population. Despite these points, emphasis was placed upon the controls and benefits to be gained by granting permission themselves instead of the DOE doing so on appeal. Concern about traffic congestion was countered with the promise of off-site road improvements to be financed by the developers.

Only one councillor argued that the committee ought to back up its expressed opposition with action, by refusing the application. The two senior councillors, however, who had decided to accept the inevitable at an early stage, stressed the positive aspects of the application, together with the inability of the district council to prevent the development going ahead within the near future. Refusal would only be a temporary reprieve, resulting in a loss of control over the terms of the outline consent and reducing the potential for planning gain. Therefore despite the many statements regretting the development, the application was granted conditional permission subject to the completion of the legal agreements. Only one Liberal councillor voted against the consent proposal; the other committee members either saw no point in a refusal or else simply followed the line adopted by their senior Conservative colleagues.

Between the meetings

The period between the two committee meetings saw a dispute between the district and county councils over the timing of the Heseltown site releases. Under section 86 of the 1980 Local Government and Planning Act, district planning authorities must consult with the county council over applications which have implications for the structure plan. The northern application had been submitted soon after the announcement of the 8,000 houses option, and the district committee consideration occurred before the county had published their strategic guidance for all the chosen areas. This guidance included the postponement of the date before which no land was to be released: from the beginning of 1983 to the beginning of 1984. If the two sites were released before 1984, they would represent 'material departures prejudicing the implementation of the provisions of the Structure Plan'. The county planning subcommittee therefore accepted the advice of their officers and lodged strategic objections to both applications.

The planning schedule adopted by the county council for the 8,000 houses maintained a five-year supply of land within Central Berkshire as a whole. Circular 9/80 on the other hand stipulated that a five-year supply was required on a district basis. The district officers therefore contested the strategic objection on the grounds that the county's phasing took no account of their own need to release land before 1984. To circumvent the objection they contacted the agents/consultants acting for the two developers, and suggested additional agreements to prevent any development occurring before the end of the year. Since neither applicant was intent on quickly developing their site, and the other Section 52 agreements would in any event take months to complete, they agreed to this proposal. The county was informed of this agreement and on the day following the second District Planning Committee meeting, a meeting of the Management Panel of the County Environment Committee dropped the strategic objections (acting for the planning sub-committee, whose next meeting would have been too late for the district's requirements). Following this move, the district council could grant permission (and therefore 'release' the sites for land availability purposes) as soon as the legal agreements were finalized.

The second meeting

The consideration of the southern site was similar in a number of respects to that of the previous application. The parish council again objected strongly, and in their letter to the district council accused the Planning Committee of ignoring their repeated and well-documented objections to development in the area. Their opposition, and that of the thirteen individuals who had submitted objections, was based on the same grounds as to the northern application. The thirteen letters contrasted with twenty-seven for the 1982 southern application. The officers' recommendation of consent was likewise derived from the same planning context as before, although the county council's objection (still in force at this point) was a new element. The nature of the committee discussion was however considerably different.

Having accepted the northern application at the previous meeting, the outcome of this application was never in any doubt. Although a number of councillors attacked the proposal and the DOE as before, this was intended primarily for the press and the few objectors who attended the meeting. It was important for the committee to make public their genuine resentment of the DOE's ultimate power over the decision, and be seen to grant permission only reluctantly. Local residents were subsequently to read a report on the application in local newspapers which commented, 'Ironically, many of the councillors who voted in favour of the development spoke against it.'

The 'reluctant approval' lobby was again led by the ward member, who also rejected the claims made by the parish council. Greater emphasis, however, was placed on detailed aspects of the development and of specific impacts. The submission of a development brief by the planning consultants, giving the suggested density, layout, dwelling mix and provision of services, focused attention upon the details. The use of this brief, which formed no part of the (outline) application and was not binding in any way, was indicative of the more sophisticated negotiating style of the consultants. The inclusion of a shop, a doctor's surgery and a number of units of sheltered accommodation showed the committee that benefits could result from the development. These features, together with the 'village green' indicated on the layout, were welcomed by councillors, and provided them with planning gains which could be used in defence of their reluctant permission. After a period of discussion over the viability of rural shops, the committee added a further condition upon the consent, to reserve land for two shop units rather than one. Legal agreements would secure satisfactory open space provision, a joint roundabout for both developments, and a contribution towards off-site road improvements. When the decision was taken on the application, only one councillor voted for refusal on the basis of the perceived inability of the village services to cope with the overall Heseltown quota.

Newhurst Nursery

Outline permission is sought

Newhurst Nursery was a seventeen-acre owner-occupied property in the parish of Warfield, just north of Bracknell. For some years the nursery with its ancillary greenhouses had fallen into decay and was something of an eyesore since part of the site fronted on to a bend in Warfield Street. Warfield Street itself was lined for most of its length by dwellings, the majority of which were bungalows or chalet bungalows.

In September 1980, the nursery owner, through his agent, submitted an outline planning application to develop ten, four-bedroomed detached dwellings on two and a half acres of the site. A development of this size necessitated the construction of an access road and the creation of a cul-de-sac, thus producing a development of some depth and contrasting with the linear development along the street as a whole. The agents of the applicant did have some pre-application discussions with the planning authority but the officers, realizing the complexity of the issues involved, could offer little guidance and suggested that the applicant should try a proposal to 'test the water'.

Responses

Following their usual procedure, the officers considered the application at their fortnightly pre-plans meeting. These meetings, which involve the deputy CPO, the Principal Development Control Officer and area officers, discuss all incoming applications very briefly to give an indication of their views on particular proposals. Generally, the question posed is 'do we agree with the principle of the development?' and in the case of the derelict nurseries, a question was also posed as to 'what else can we do with the site?' The decision was finely balanced and the initial reaction ensuing was negative. This reaction was publicized as usual through the fortnightly listing of planning applications which is sent to parish councils and other groups and individuals who wish to be consulted.

The initial reaction is never binding, and after a site visit and further consideration, the area officer with the support of his superiors reversed the initial reaction and recommended approval. This was not a major about-face but rather a delicate tipping of the balance.

It was feared that if the applicant went to appeal an inspector might easily grant permission. Moreover, it was difficult to press for a smaller (infill type) development because the county surveyor had indicated that since the site was on a bend in the road more than one access point would be considered as dangerous. Thus a small frontage development would be unacceptable whereas the alternative, providing one access road, would inevitably open up the site for a larger development. In the event, the officers opted for the latter classifying it as rounding-off in terms of structure plan policies. Moreover, the proposal was also justified on the grounds that the deterioration of the site would be halted and the local environment thereby improved (and again structure plan policies were cited). So the applicant had benefited principally from the bend in the road and partly from the condition of the site, and obtained officers' approval for a ten-dwelling proposal.

Acting in response to the original officers' reaction, the parish council in their comment on the application briefly stated that they 'agree[d] refusal'. According to the planning case file no other objections were lodged until two days before the district Development Committee meeting.

However, prior to submission, the applicant had approached the local ward member who had stated that she would oppose the application. She had suggested that she would only be amenable to a slightly stepped-back frontage development of five dwellings. Her consistently held views about the village were that no large-scale development should be permitted and that development should be limited to infilling and house extensions. She feared that the entire seventeen-acre property and not just the two-and-a-half-acre site might be opened up for development.

Realizing that the officers' recommendation would not go through unopposed, the applicant decided to convene a site meeting. He wrote to all the members of the development committee inviting them to visit the site one Saturday morning in January 1981. The meeting was unofficial and did not involve the officers of the district. The response was good and some ten members of the sixteen-strong committee attended. The local ward member turned up considering that the other members would do no more than refuse the application. However, the applicant, suitably attired in a sack apron, appealed to the members that he was a poor man and that he had striven unsuccessfully to make his nursery viable. He could not do so and therefore had decided to sell. It seems that the small businessmen amongst the Conservatives were impressed by the tale and the Labour members were happy to see the provision of more housing. The members took away positive views of the development although at that stage they could not of course issue a decision.

On the Sunday before the Thursday Development Committee meeting the parish council chairman was informed by the local district member that the officers' recommendation was for approval and was not in line with the initial reaction. A meeting of the parish council planning sub-committee was hastily convened early in the week. Reaction was strong against this application because of the context set by other applications for Green Belt development in the area. A letter was composed which complained about the lack of consultation over the change in the reaction. The parish council still objected to the application but in the last paragraph 'an off-the-cuff remark' was included which suggested that twenty low-cost dwellings which would probably be in the first-time buyer's range would be preferable to ten up-market dwellings. This comment admirably reflects the conflict of views in the parish: while many recognized the need for low-cost dwellings, others (including the parish council) were fearful of articulating such views for fear of being engulfed by a very large number of dwellings. Low-cost dwellings were desired by some sections of the village in the hope that the demographic base of the village would be broadened and that the children of the newcomers would help to keep the local school open.

The committee decides

At committee, the local ward member spoke against the proposal but she was not supported by many of the other Conservatives who voted for the proposal. One of the leading Labour members was reported to have taken great delight at the parish council letter – he had always favoured development in Warfield and felt that at last the parish council had 'woken up'.

In the event, the application was approved and the applicant could have been in little doubt that the committee might accept an even higher density of development. However, along with the approval notice, the officers had recommended and the committee agreed, to send an 'informative'. The informative stated that the 'design and size of units [were] to reflect the visual character of the immediate locality' and was included to try to dissuade the applicant from submitting a proposal involving very large houses, which were hinted at on the outline proposal.

After the decision had been issued by the committee, local opposition to the plan became more vocal. The local amenity society wrote to the district council claiming that ten new dwellings would upset Warfield, and in fact a neighbour of the development took the issue of the reaction and the handling of the application to the Ombudsman. His case was dismissed.

A full application is submitted

So, against this background of bitterness and recrimination, a full application was submitted six weeks after the outline approval. And sure enough, the applicant did come back to ask for more – a total of twenty-seven dwellings. A twenty-dwelling proposal in line with the parish council comments had been considered by the applicant but was dropped when it was realized that the construction of twenty first-time buyer's dwellings on such a site would not be viable. Twenty-seven first-time buyer's dwellings would be viable but the planning authority made it quite clear that this was too many. The applicant subsequently reduced the figure to eighteen houses by amended plans.

By this time, the parish council had responded and backtracked on its earlier letter saying that the original letter had not had the backing of the full parish council and that while they were aware of the problem of low-cost accommodation availability, they considered that a development of more than ten dwellings would be unacceptable.

Meanwhile, the new application succeeded in arousing considerable village opposition which was articulated in fourteen letters of objection and a petition with 203 signatures. The objections mainly centred around the density and character of the proposed development and the traffic hazard it might produce, together with a general lack of facilities to support the development. Some letters even boldly stated that their own properties would be devalued. Against this, there was a single letter in support of the application.

Representations were also made to the local member who said that there was little she could do to fight the principle of development, outline permission having been granted, but that she would fight very hard to ensure that the dwellings were bungalows and not houses. Bungalows, she felt, would blend in with the street better and avoid the problems of overlooking about which some neighbours were concerned. Although the applicant contacted her on several occasions, she refused to discuss the matter with him giving her stock response to developers: 'It is for the officers not the members to undertake negotiations with applicants.'

In defence of the application, the applicant's agent wrote to the planning authority emphasizing the inevitability of a high-price for houses on a ten-dwelling site and claiming that the construction of anything less than eighteen dwellings would not be commercially viable. The letter clearly exploited the parish council's earlier letter, the views of the Labour members to housing provision, and the attitudes of the Conservatives to profit motives.

The application is rejected

Realizing the sensitivity of the whole issue, the planning officers' report painstakingly responded to the claims of the objectors and set forward their views – that thirteen units might be appropriate and that these should be a mixture of dwelling types and sizes. They restated the point that frontage development was unacceptable to the county surveyor but that the cul-de-sac development might be acceptable if the dwellings to the front of the site were larger than those to the rear. Claiming that the current application did not concur with structure plan policies which required that proposals should reflect the physical and visual characteristics of their surrounds, the application was recommended to be refused. The committee duly refused the proposal by the very narrow majority of one – a split which seems to have reflected the Conservative/opposition parties split. Furthermore, the committee would not agree to the suggestion of the officers to include an informative implying that thirteen dwellings might be appropriate.

After the applicant had received the initial outline permission, he had instructed his solicitors and agents to begin marketing the site. As is the case when land is fed to agents, the land was probably offered to a few developers and then to a wider range. One developer replied that their company was interested in the site but felt that it still had serious problems – drainage had to be laid across other people's property. The developer therefore stated that they would only make an offer on the basis of a conditional contract, i.e. the sale would only go ahead when drainage and detailed planning permission had been granted.

Whether offers from other developers were also made is unknown but it is thought that any who did respond must also have expressed concern about the drainage and planning matters. In any event, takers seem to have been few and after hearing nothing for four or five months the medium-sized developer was approached with a view to drawing up the proposed conditional contract. As a 'gesture of goodwill', the developer also agreed to research and fund an appeal by written representation over the second application. However, it should be noted that it was the original applicant and not the developer who was pressing for the appeal. The developer, who concentrated on up-market dwellings, considered that the site was inappropriate for a large number of dwellings; the developer would have been content with the ten-dwelling outline permission.

The appeal

None the less, the company undertook the appeal and were unsurprised by the inspector's decision to dismiss it. Even in the period following Circulars 9/80 and 22/80, which had done so much to change the planning system, the DOE was unprepared to allow such an appeal. In dismissing the appeal, the inspector considered that the proposed development would amount to extending the built-up area of the village into the countryside, would effectively create a new street within the village and would be detrimental to the traditional character of a linear-shaped English village which Warfield seemed to have preserved. The inspector also made the pointed remark that the eighteen-dwelling application bore no relation to the original ten-dwelling outline permission.

A new application

Vindicated in their views, the company submitted a 'reserved matter' application based on the original outline proposal in September 1982. Although the company had not yet bought the site and were not incurring interest rate charges, there was some urgency to feed the site through the land bank cycle because they wanted it to be part of their 1983 development programme.

A previous committee informative had strongly indicated that the development should blend in with Warfield Street. Indeed, the developer had from the beginning always considered some form of mixed development and the proposal submitted was for six houses and four bungalows. Bungalows are generally less profitable, but in the hope of gaining a speedy approval the developer conceded one or two more bungalows than might otherwise have been the case. Although the developers did approach the local district member she would not consider speaking to them in any detail.

At first this proposal met with parish council approval but the local amenity society, the Warfield and District Amenity Society (WADAS), one of whose leading members was the local district councillor, objected to the proposal and forwarded a scheme which included two houses (on the frontage) and eight bungalows.

Representations were made by local residents to the parish council (some say they were encouraged by the local district member) and the parish council decided to call a planning subcommittee meeting to which interested parties could come. The council heard their views and the local district member claimed that the officers would support an application for ten bungalows. Somewhat taken aback by this revelation, the parish council requested a meeting with the district officers. At this meeting the officers clarified the situation by stating that they could only react to development proposals and that in doing so on this site, a prime consideration was whether or not their decision would stand up to the test of appeal. The parish council felt reinforced in their view that ten bungalows were not a viable proposition.

In presenting the application to committee in January 1983, the officers recommended approval for the applicants' scheme.

Another refusal

However, the scheme was blocked by the committee who instructed the officers to issue an informative along with the refusal notice, stating that a new proposal in line with that suggested by the petitioners would be looked upon favourably by the committee.

Another application

Within a week of receiving the refusal notice, a new application was submitted which went a little of the way to meet the objector's demands – this application was for five houses and five bungalows. Within another few days, the applicant put the planning authority under pressure by threatening to take the previous decision to appeal with a view to both winning the appeal and gaining costs.

Meanwhile activity within the village was hotting up and a petition of thirty-three signatures was drawn up in *support* of the application. Letters in support of applications are very rare. While the local district councillor claims that this supportive petition was instigated by the applicant, a leading Labour member insists that it was initiated by three people who genuinely wanted more low-cost dwellings for Warfield. Two days later another petition (of fifteen signatures) appeared from the most affected residents objecting to the proposal. As a planning officer at the time noted, one person had signed both petitions. Furthermore, a local residents' plan volunteered yet another possible proposal – for one house and nine bungalows.

This time the parish council stayed their hand and arranged to report their comments verbally at the Development Committee. In doing so, they suggested that the development should be predominantly of bungalows and should therefore consist of four houses and six bungalows.

In presenting their report to committee, the officers did not forward a recommendation but took the highly unusual step of stating, 'The committee's instructions are requested.' The developers were somewhat peeved by this action even though the report explicitly stated that the proposal was 'generally acceptable'. The officers acknowledged that the informative attached to the decision for the third application had not been adhered to but considered that there was little chance of achieving this through further negotiations.

The committee decides . . . again

The committee, yet again voting on the issue, decided by a narrow majority to approve the application.

Development has begun but the rumours about the site are rife. A road has been constructed to allow access to the rear of the nursery site. Some claim that this will open the way for further development but the planning officers felt obliged to allow the construction of the road because no other access to the land could be proved to exist. Nevertheless, the planning officers are quite clear in their minds that they will not recommend approval for any development on the rear of the site.

Emerging tensions . . . again but finally

In Figure 2.4 (p. 36), a number of relationships between agents were outlined. In this, the final section of the last chapter, we will reconsider most of the inter-relations by focusing on two main tensions.

Private sector growth and public sector retrenchment

In post-war Britain as in other advanced capitalist countries the state has become a major provider of goods and services. Much work in recent years has sought to read off capital's interests from the nature and level of this state intervention (e.g. Harvey, 1982). However the mechanical nature of this functionalist argument fails to note at an empirical level the changing demarcation between market and public provision while at a more theoretical level it ignores the existence of the tension between the interests of capital seeking socialization of costs and the political representatives at both local and central government seeking to reduce rates and taxes. Moreover the context of state intervention has varied over time. Since the early 1970s economic crisis has replaced economic boom and in Britain the recession has had two main consequences. First, fiscal constraints have been placed on government actions as revenues have been limited while expenditures have increased. Second, a fertile climate has been created for the development of monetarist and social market ideologies which stress the dangers of high levels of government involvement and expenditure (Hayek, 1982; Friedman and Friedman, 1980). These ideas have been readily adopted by important elements of the Conservative party and constitute the intellectual basis of Thatcher's policies and to a lesser extent of the previous Labour administration (Hall and Jacques, 1983). The first constraint reduces the state's ability to pay for services while the political consequences of the latter reduces its willingness to pay for them.

Since the mid-1970s the control of public expenditure has been a major government aim in Britain. Because local government expenditure accounts for almost a quarter of total public expenditure, central government has sought to limit local authority spending. It has been able to do so because between a half and two-thirds of local authorities' income derives from central government grants. Central government strategy throughout the 1970s and into the 1980s has been to reduce the size of this grant. Thus in the county of Berkshire the block grant in 1981/2 from central government was £66.6m. but in 1982/3 it was only £45.7m. The reduction has come at a time when other costs are rising. Extra burden is also faced by growth areas due to the provision of extra facilities. Table 7.1, for example, gives a rough and general indication by the county planners of the type and level of spending for the different public services necessary to complement population growth. Estimates in the structure plan suggest that the provision of public services for a population of 643,500 in 1986 would involve extra capital expenditure of £60.3m. (1977 prices) and a 3.4 per cent annual average growth in revenue expenditure. The costs of growth are high. In the long term the new developments produce extra rates and hence income. But in the short to medium term there is disparity in the costs and income as the necessary infrastructure has to be provided preceding and during construction before rates are paid. The gap between costs and revenue is not being bridged by the central state and if anything the gap will widen as a consequence of the government's current fiscal policy. Part of the gap can be met from extra rates (local property taxes). The rate precept of district councils in Berkshire increased from 112p. in the £ in 1981/2 to 142.8p. in 1982/3: an increase of 27.5 per cent. However, there are political limits to this magnitude of rate increase. Local electors are particularly sensitive to rate increases, and vote accordingly. In the 1983 local elections all the major political parties fought the election in Berkshire on a platform of no rate increases.

TABLE 7.1: Estimates of costs of alternative levels of population growth (1975 prices)

Service	*Costs (£'000s) of population growth of 50,000*	*Costs (£'000s) of population growth of 150,000*
Education	10,100	23,600
Libraries	700	1,700
Social services	4,800	10,600
Protective services	600	1,700
Other	1,100	3,300
Total	17,300	40,900

Source: Berkshire County Council, 1976

In Central Berkshire, therefore, private growth has been occurring at a time of redirection in public expenditure.

Because much of the physical and socio-cultural infrastructure which enables residential development to take place (e.g. sewerage, roads, education and recreational facilities) are provided by the local authorities and since the rapid scale of such developments in Central Berkshire is occurring at a time of fiscal constraints there have been a number of consequences and two main responses. The consequences are obvious in the increasing road traffic congestion and the extra pressure on residential and recreational facilities at certain locations. Although a relatively affluent area, the experience of growth is reminiscent of Galbraith's (1958) comments on private affluence and public squalor.

Response of the local planning authorities

The main response of the planning authorities to the situation has been to seek *planning gain*. This involves agreements between local planning authorities and developers who are required to provide certain benefits to the community as a condition of obtaining planning permission. As Table 7.2 shows, planning gain agreements have been increasing.

Three types of planning gain agreement can be identified. First there are *ameliorative gains*, where developers carry out or contribute towards improvements in infrastructure which are necessitated by their own development. As an example, two ameliorative gains were obtained from the planning consent for 500 of the additional 8,000 houses which were located in a village near Reading. The developers agreed to provide extra drainage capacity, including a new pumping station, and to carry out/

finance improvements to the road connecting the village with Reading. This was already inadequate for the existing traffic load,

TABLE 7.2: Planning applications and planning gain agreements in Wokingham District Council

	Total planning application	*Nos of agreements*	*% of applications with agreements*
1974	1694	6	0.35
1975	1908	2	0.10
1976	1827	13	0.71
1977	1649	18	1.09
1978	2175	21	0.96
1979	2363	20	0.84
1980	2379	31	1.30
1981 (first six months)	1614	22	1.36

Source: Henry (1982)

and in the view of the Highway Authority could not cope with the demands of a further 500 households.

The second type is *compensatory planning gain* where the planning authority seeks some benefit to the community at large from the granting of planning permission to a developer. In the words of the chairperson of a Central Berkshire planning committee, 'if an applicant/developer is obtaining benefit from a change or intensification of use on a site, then he is the right person to contribute to the community.' Gains of this type include leisure and community facilities, footpaths, riverside walkways and footbridges, and amenity/recreational areas (see Block 25). The insistence of planning authorities that residential and retail elements be included within office schemes, and demands for either full parking facilities or commuted payments to the local authority for public parking provision, can also be considered as compensatory gains.

The third category can be described as *planning bartering*, and has often been critically viewed as selling planning permission. It usually relates to larger developments where the local authority actively seeks an objective and is prepared to grant permission to the developers in order to obtain the construction or financing of that objective. Two examples of this form of planning gain have recently received consent within Reading. One concerns a relief road for the busy A33 through the south of town. This road will be constructed by a development company as part of a large-scale industrial scheme alongside the M4 motorway. Without the industrial

Block 25 Planning gain 1 – Seeking community facilities

Leisure plan magnet for developers

DEVELOPERS are queuing up to be part of Reading Borough Council's plans to build a huge leisure complex on the Richfield Avenue, Reading, site, along with some industrial/commercial units and a 150-bedroomed hotel.

A total of 15 interested parties – companies, individuals and groups of professionals – have held discussions with the council's Chief Executive, Mr Harry Tee, about the proposals.

The plan is that a developer will be granted a lease of up to 125 years on the sites of the industrial/commercial units and the hotel, and in return will, at no cost to the council, provide a leisure complex to include:

- Sports hall and courts.
- Leisure pool.
- Open air children's facilities.
- Pavilion accommodation including changing rooms, toilets, showers and bar facilities.
- Approximately 20 to 25 acres of laid out playing fields including 'all weather' pitches.

The developer will also be required to provide car parking and pedestrian access and to carry out highway improvements and any repairs to the riverbank.

Details of the proposals are contained in a report by Mr Tee which was being considered by the council's leisure committee yesterday (Thursday).

The Chief Executive recommended that one developer should tender for the whole scheme and that he should provide details of the ground rent to be paid to the council, the layout and outline design of the commercial/Industrial units, hotel and leisure complexes, car parking and landscaping and the method of financing.

Mr Tee suggested that a small sub-committee of the policy committee should be appointed to settle the list of developers to be invited to submit schemes.

He estimated that work could begin on the 50-acre site by the middle of 1984.

Members have already approved the principle of a centre being established in Reading but had to decide whether it should be at Richfield Avenue or on the Whitley Recreation Ground, to be developed in conjunction with the ice rink.

development the relief road would be postponed until Berkshire County Council had the finances available and had secured the purchase of the land – a number of years (see Block 26).

The other example concerns the railway station site within Reading. As part of a complex redevelopment a new and larger station concourse would be built together with much needed

Block 26 Planning gain 2 – Tactics and pressure

COMPANY WANT PLANNING PERMISSION FOR DEVELOPMENT

Threat to withdraw relief road offer

A LOCAL company which is willing to build a major new road in Reading at no cost to the ratepayers, has threatened to withdraw the offer unless planning permission is granted for a large industrial development near the Courage Brewery.

Rickworth Securities have been refused permission by Wokingham District Council to develop 64 acres of land which they own alongside the M4. The company, formerly the Lesser Property Group and Caversham Bridge Holdings, wants to provide one million square feet of factories and warehouses which they say will create 2,000 new jobs.

They also want to build the second stage of the A33 relief road from the M4 to Rose Kiln Lane at a cost of £3 million and contribute an extra £½ million towards improvements to Junction 11 of the motorway.

But the application was refused because it contravenes the Central Berkshire Structure Plan which makes no provision for industrial use on the site.

Now Rickworth Securities have said they may go to appeal and because of the cost involved the company might withdraw its offer to build the A33 relief road.

Managing director of Caversham Bridge Holdings, Mr Tony Butler, said: 'The longer we wait the higher the costs are going. It's a question of time. We have got to do it quickly.'

He added: 'We have never said this relief road would take all the traffic off the Basingstoke Road, but it would make a major contribution to easing congestion. Traffic wanting to go westwards from the South Reading area would then use the new road taking a good deal of traffic out of Elgar Road.'

Mr Butler said the road would also open up the Reading Borough Council owned Smallmead Tip site which could then be developed.

Reading Council has supported the plan but Wokingham District Council is the planning authority for the area and it rejected the application about two months ago.

Because it contravened the Central Berkshire Structure Plan the County Council was consulted and it also raised objections. Councillors said there was more than enough provision in the Structure Plan for industrial development and that the application would prejudice the review of the Structure Plan currently underway.

Rickworth Securities are still in regular contact with Wokingham's planning officers in the hope that some agreement may be reached over the development. Mr Butler said: 'I'm sure that commonsense will prevail in the end. If it went to appeal it could take another year but Wokingham have said they will keep talking.'

Originally, Rickworth Securities offered to build the relief road from the M4 all the way to Berkeley Avenue. Now the County Council has said it will build part of the road – from Berkeley Avenue to Rose Kiln Lane – so the company has instead offered to contribute £½ million towards improvement to Junction 11 of the motorway.

Mr Butler said: 'Reading is being advertised as a place to relocate to, but what it lacks is good roads to the motorway network. With increasing traffic Reading needs not only the widening of existing highways but the building of new ones.'

Source: *Reading Chronicle*, 21 October 1983

additional parking and a large area of speculative office development. The policies of the borough council towards office development have been generally restrictive, but the pressing need to replace the inadequate station has persuaded the Tory-controlled council to permit more office space than originally intended.

The pursuit of planning gain has a number of consequences. In seeking to maximize planning gain the local planning authorities have tended to look favourably on large site schemes involving the large builder-developers where planning gain agreements can be more easily reached and implemented. The decision to put 4,000 Heseltown houses in north east Bracknell for example was partly made on the criterion that planning gain, particularly over road and sewerage costs, could be effectively achieved on such a large site.

There are thus marked redistributional consequences within the housebuilding sector. It is only the large housebuilders which can successfully operate on large greenfield sites. As we have seen, these sites entail considerable infrastructure and planning gain costs which only the large builders can contemplate. The large sites have made minimal or negative impact upon the small housebuilders who continue to develop the small infill sites or urban areas. But the local medium-sized builders are caught in a dilemma: whether to compete with the large housebuilders on the large sites and risk failure because of their lack of resources, limited negotiating expertise and lack of proven ability to produce and sell at relatively low cost, or to continue on other sites with their older practices and risk being squeezed out of the land market.

The response of the community

Planning gain has been most vigorously pursued with respect to those costs previously directly borne by the local authorities. It has been much more difficult to obtain socio-cultural facilities such as large amenity areas, widespread tree planting, community centres, etc., which are neither a statutory nor a technical necessity. The failure of either planning gain or the local authorities directly to provide such facilities has provoked a second response, that of residents' associations, community groups, and parish councils seeking to improve their own local environments. With reference to the resident groups, the response of the getters and combined stopper-getter groups have been twofold. On the one hand they have sought to redirect public and private resource endowments from either developers or the local authorities. Most success has been achieved with the local authorities and parish councils since they have more of a duty and a need to meet local demands than developers. However, given the fiscal constraints, successes have been slight. More successful have been the self-help schemes where groups have responded by either topping up or directly providing socio-cultural services from their own resources. Action has varied from providing books for local schools to the construction and management of community centres. But the self-help schemes have introduced new sources of inequalities since communities vary in wealth and the ability to transform wealth into resources.

The existence of planning gain should force us to reconsider our views of the relationship between state and capital. Crude Marxist views see a mechanistic relationship with the state always underwriting the costs of capital. Although more recent interpretations are aware of recommodification and privatization and note the contingent division between the public and private sector, the underlying hypothesis is that the division is always set to aid capital. Planning gain highlights the important case of the state passing costs, at least in the first instance, on to capital. The actual cost of planning gain is in the first instance borne by the developers but this is passed on in two directions; in the sale or leasing of the sites and buildings, and on to the landowners. Most developers purchase land through option agreements, whereby the landowners receive an agreed percentage of its market value (varying between 80 and 95 per cent) with the infrastructure costs deducted. Although these option agreements require the developer to limit these costs to essential expenditure, in practice it is often possible to add planning gain costs to the infrastructure bill. Where the costs of planning gain cannot be transferred to the landowners, then they are added to the house prices or office/industrial rents. Planning gain thus acts both as a form of hidden tax on incoming users and purchasers and a development charge on landowners.

Local interests and national opportunity

Against the background of high development pressure the state itself, or more precisely the planning system, becomes the arena for the articulation of the various interests involved in the production and consumption of the built environment. The resultant tensions are evident in the conflict between local interests and national concerns and expressed in central–local planning relations.

In a prosperous area such as Central Berkshire the policies of employment generation and civic boosterism have restricted political purchase. Most people in the area have jobs. To be sure there is unemployment but the dominant voice is that of the no-growth lobby. The most articulated concern is with the maintenance of the exclusivity of local social environments and the conservation of the physical environment. Environmental issues are run together with concerns over house prices and social status. The no-growth lobby is evident in two forms. On the one hand there is the formal political representation in council chambers. The majority political view at both county and district level is that pressures for office and housing developments should be resisted. The Chief Planning Officer of the county has noted that the prevailing view of the political members of Berkshire County Council is that pressures for office and housing development should be resisted (Stoddart, 1983), and a similar situation exists in most of the districts. On the other hand there are the large number of stoppers and stopper-getter residents' groups who individually and collectively limit the area of institutional and physical space open to the local planning authorities seeking to manage growth. The parish councils have an ambivalent position between formal and informal representation. Although they are statutorily defined and entitled to comment on planning decisions, thus forming elements of the state apparatus, they are very receptive to local pressure and thus constitute a significant channel for local opinion. In Central Berkshire the dominant theme in parish planning debates is for the need to restrict and deflect residential development.

However, despite all this local interest, ultimate control rests with central government, or to be more precise the Secretary of State for the Environment. Throughout this book we have made reference to the structure plan preparation and modification in Central Berkshire. The structure plan submitted in June 1978 suggested that 32,000 houses were to be built in Central Berkshire up to 1986. Formulated in a local atmosphere of anti-growth this plan revealed the reality of central government control. If the structure plan had resisted development pressure, as suggested by most of the local groups involved in the public participation exercise, then central government, ever mindful of builders' interests and 'national economic opportunity' in one of Britain's few growth areas, would simply have allowed development on appeal. Since this would have meant local planning authorities losing control over the type, location and extent of development even the unmodified structure plan represented a willingness to accept growth. The extra 8,000 home modification made by the Secretary of State in 1980, the Heseltown amendment, is only the most visible sign of central government influence. The fact that land was to be released to accommodate an extra 100,000 people by 1986 shows the potency of the developers' lobbies and central government control. The government has not been oblivious, however, to the possible repercussions of conflict with the local concerns of its political heartland. It is no coincidence that the draft circulars on land availability and greenbelts, which suggested more land release, lay on the minister's desk for months and were not released until after the 1983 general election. And in the context of this case study, it is important to note that the whole Heseltown issue was only brought to the surface in Central Berkshire by opposition parties eager to embarrass local Tories with the actions of a Conservative minister.

The apparent tension between national interest and local concern is in reality a conflict between developers' interests and macro-economic objectives on the one hand and the articulated concerns of local residents on the other. Building capital has found more purchase for achieving their goals from a central government mindful of maximizing growth and encouraging an investment-led economic recovery. Local interests have achieved most purchase through local representation. Given the ultimate power of central governments in land use planning matters, the example of the Central Berkshire Structure Plan shows how development interests have won over those of local residents.

In this context elected members in the local authorities have been placed in a difficult position. On the one hand they are forced to administer a planning system which is ultimately pro-growth, but on the other hand most of them represent constituencies where anti-growth is the dominant opinion and the safest electoral posture. The local politicians, especially at the district level where most development control decisions are made, have to operate a national system of planning which is locally unpopular. Their response has been twofold. First, they blame central government. However, there is limited political mileage in that strategy. Second, they seek to gain some form of planning gain for the community. In one sense planning gain legitimizes the position of elected members. Councillors can show their positive involvement in the planning process obtaining benefits for the local community. The more the councillors can obtain from developers the greater the self-justification.

In the introduction we posed the issue of how the state resolves the dual pressures of maintaining capital accumulation and ensuring social legitimation. In this particular case the demands of developers have been mostly met in Central Berkshire by imposition of 'national concerns over local interests'. The local state has been more responsive to the no-growth lobby. This would seem to confirm Saunders' point about the division between the central and local levels of government overlying the competing claims of accumulation and legitimation (Saunders, 1980). Given the nature of British land use planning the local planning authorities have had to, albeit under some protest, accept central government directives, through circulars, structure plan modifications and the threat of development on appeal. However, the position of the elected representatives has to some extent been legitimated by the successful pursuit of planning gain; this gain neatly meshes local political interests with central government macro-economic policies of reducing public expenditure.

Appendix 1 The Housebuilders

Generating the housebuilder sample

The term housebuilder is used in this book to refer to an enterprise which buys land and builds houses thereon for sale on the open market. There are a number of possible sources of information which can be used to generate a sample of housebuilders operating in Central Berkshire over the period 1974 to 1982. The Yellow Pages parts of local telephone directories have a section on Building Contractors (Builders in the latest directories) but this source was found to include a vast number of non-builders and excluded the larger companies with sites in Central Berkshire but with offices elsewhere. The builder-developer listing in the *House Builders' Federation (HBF) Year Book* includes only larger enterprises (only seventy-three were listed for the whole of Berkshire) and categorizes only the main office of national/regional operators. The National House Building Council (NHBC) register of builders and developers also suffered from the same drawback.

The best source remaining was the planning register. To undertake development, it is necessary to obtain planning permission. The applicant may be the owner of the land or merely have an interest in it, and, in nearly all cases, this means that the applicant is the landowner and/or a builder-developer.

Berkshire County Council operates a Development Control Information and Monitoring System (DCIMS) from which it is possible to obtain a computer print-out of the salient details of residential planning applications for the three districts of Central Berkshire: Newbury, Wokingham and Bracknell; Reading Borough is not included in this system, but information from Reading Borough's planning register was compiled separately and became part of the sampling frame. From 2,644 successful residential planning applications, 1,593 different names were listed, only some of whom could be builder-developers. By including only multiple applicants a sampling frame of 175 was established. The original aim of undertaking a postal questionnaire proved inadvisable because of the difficulties in obtaining addresses for each name in the sampling frame and because of the limited range of information which would result. Instead forty housebuilders were randomly selected from the sampling frame and interviewed. Very few firms (an additional four) declined an interview. The range of firms interviewed was large and included local, non-local and national companies; individual, family and corporate firms; and specialist and non-specialist housebuilders.

The interview

The interviews conducted with those firms ranged in duration from one to just over three hours and a wide range of information was obtained. Occasionally, it proved difficult to contact one person who could supply all the relevant information and sometimes it was necessary to interview more than one employee of the very large companies. The quality of the information supplied varied considerably and answers were not available for all the questions. This, however, is inevitable with all detailed questionnaires; ultimately the result was very satisfactory. The interview was designed to elicit certain precise facts, to obtain other information by the means of precoded responses, and to encourage general discussion on some topics.

The interview began with some general background information about the interviewee and the firm. These questions elicited considerable information and, where available, were supplemented by company reports. Inquiries of institutional membership and holding of office provided basic information and, in a few instances, provided some interesting detail from active members. Few held public office of any sort.

Information on employees was readily available from most but complications arose in distinguishing those involved in housebuilding from those involved in other company enterprises. Figures for annual dwelling completions, though frequently incomplete for the full span of the survey, proved very revealing and a better indication of company dynamics than figures for annual turnover. Respondents were very forthcoming about future policy and prospects. Together this information gave a very good indication of the scale and scope of companies' interests and involvement with housebuilding.

Price and type of dwellings constructed took time to elicit and was not always comprehensive in content. None the less, it did help to distinguish certain types of housebuilder which, overall, proved very useful. A seemingly innocent question concerning the use of an architect also gave good insights of the style and approach of the company.

Surprisingly, nearly all respondents were forthcoming about their sources of finance, though precise figures were seldom reliable simply because of the vast range of accounting practices of companies.

With the exception of a few of the larger developers (who simply could not, rather than would not, cooperate), information about land banks was freely given.

Queries about land acquisition policies proved fruitful and respondents had little difficulty in responding in the prescribed manner.

A series of questions requesting information about the links of companies with estate agents, building societies and consumers served to highlight how developers related to various outside agencies and their approach to marketing. Inquiries of links with other builder-developers provided polarized responses indicating the competition and cooperation of companies of various sizes.

A section on links with planning authorities tried to elicit some basic themes. In so far as these went, the questions were successful but proved difficult to code on a comprehensive basis. The item on possible changes to the planning system produced much animated discussion and revealed a great deal about the style of various companies. Following on from this, an attempt was made to discover those developers who had regular links with community groups.

Finally, a series of questions on 'Heseltown' (the then imminent allocation of extra land for residential development) brought out some of the ways in which developers approached what often erroneously appears to be a unique situation. As always, informal discussion after the standard interview provided some very useful and indicative background information.

Classification of the sample

To gain a better understanding of the building industry, various indices have been adopted to classify enterprises by size (see, for instance, Ball, 1983 and Bather, 1973). Size is a useful basis for categorization since it strongly influences modes of operation and structure of organization. Inevitably there are anomalies, and there are better ways to classify enterprises for specific purposes, but in general, size is a remarkably good indicator of the many facets of a firm.

But on what basis can a classification by size proceed? Traditionally, three main measures have been utilized: numbers of employees, numbers of dwelling completions, and annual turnover. None of these is easy to manipulate and in themselves can provide only a partial impression of the nature of an enterprise.

The labour component

Since the end of the last war, the building industry has undergone vast shifts in employment patterns. Employees required by a builder-developer range from the administrative, secretarial and sales staff to site labour. Viewed in numerical terms the site component is by far the most important but it is not really feasible to obtain meaningful figures for the numbers of site employees. Extraordinarily few are now employed on a PAYE basis; some operate on a direct but self-employed basis but the bulk are indirectly employed via contracting. As one employer commented, 'My job demands that I do not think of the site workforce as people – I must view site work in terms of purchase of completed jobs.'

The notorious fluctuations of the construction industry over seasonal and more long-term cycles have encouraged most builders either to shed or never to engage directly employed site labour. While this may lead to higher rates for a job of work in the short term, the longer-term benefits for the builders are considerable. Housebuilders are able to cease or slow their building rates and reduce their running costs in time of slumps until the market improves. To include site labour in an assessment of business size would be extremely misleading because of fluctuations, the nature of subcontracting, and the overlap of labour inputs to housebuilding and other construction sectors.

Categorization by annual turnover and level of profit

Annual turnover (i.e. summation of selling prices of all dwellings) seems the most appropriate way of measuring the size of a firm. However, two main problems are encountered.

First, not all interviewees could or were willing to give figures on annual turnover. It was not always possible to separate turnover from housebuilding and other enterprises. Moreover, strong arguments can be made in support of a classification based on housebuilding *and* total company turnover.

Second, classification by annual turnover effectively prohibits comparisons over time. Price rises during the 1970s effectively make standardization impossible.

A classification based upon profit levels would provide a fascinating insight of the housebuilding industry. Unfortunately it is impractical. Even if interviewees were willing to divulge such figures, their derivation would be highly complex and most likely not afford comparisons.

The major classification adopted

Despite its drawbacks the most suitable and illuminating categorization of housebuilders is by number of annual completions. The discussions in Chapter 3 focus upon this classification. The divisions used are of course arbitrary but Table A1.1 shows how conveniently they also relate to annual turnover. The categories were not devised so that the completion and turnover divisions should comply, but it is notable that only one builder would fall into a different category if assessed by turnover alone.

TABLE A1.1: Classifying the sample

Number in sample	*Housebuilder size*	*Annual dwelling completions*	*1981 turnover limits*
10	Very small or occasional	0–4	£0–290,000
9	Small	5–20	£290,000–£700,000
7	Medium	21–100	£1,400,000–£3,500,000
14	Large	100+	£12,000,000+

The sample in relation to the national picture

How representative is this sample from Central Berkshire of the housebuilding industry in Britain as a whole? From the planning files it was possible to estimate for both the sampling frame of 175 and the sample of forty the largest site of applicant in terms of number of dwellings. These figures are compared in Table A1.2 with information on the national distribution of housebuilders by annual output. Two things are clear. First, the sampling frame contains a higher proportion of the larger housebuilders than the national average. This is due to the fact that Central Berkshire, as a growth area, attracts the larger operators and to the decision to only include multiple applicants. Second, the interviewed sample contains a distribution even more skewed to the larger operators. While the sample provides a coverage of the range of builder size, our sample contains a higher proportion of larger operators than the national average.

TABLE A1.2: Comparison of housebuilder size distributions: the interview sample, the Central Berkshire sampling frame and the national situation

Largest site of applicant	*Interviewed firms* (Number)	(%)	*Central Berkshire sample 1974–81* (Number)	(%)	*National size distribution 1982** (%)
1–10	13	(33)	94	(54)	84
11–50	4	(10)	38	(22)	11
51–100	2	(5)	19	(11)	3
101+	21	(53)	24	(14)	2
	40		175		

*NHBC data – annual output Base = 8212

The dynamics of housebuilding firms

The suggestion that the most appropriate classification of housebuilders is by size immediately begs the question as to how frequently firms may require reclassification as their output fluctuates. The answer is seldom, but, in part, this is a function of the category divisions chosen. The divisions were not randomly selected; they derived from the idea that each category operated in an identifiably different fashion. The nature of these operations, it is suggested, imposes limitations upon output and these limitations are only effectively overcome by major changes in operation. From the information available, and it is not comprehensive, there is no evidence to suggest that there is a 'natural' growth or life cycle of a housebuilding firm which causes a transformation in size. On the contrary, it appears that (in the last two decades at least) a firm becomes involved in housebuilding at a level of output from which it will deviate little. It is not thought that this merely reflects the recession of the early 1980s; the process seems to have been in train throughout the 1970s and before.

Three major factors tie housebuilders to a certain level of output. First, there are the desires and aims of a company. Inertia is a powerful force and management is frequently wary of altering course without fully comprehending the outcome.

Second, there is the limiting factor of capital provision. Land, the housebuilder's resource, is expensive and money must be placed 'up front'. Lack of this resource imposes considerable problems in achieving a policy of growth.

Third, there is the problem of land search and acquisition. Capital needs to be invested in good land but this resource too is restricted.

Output does vary but the variation tends to occur within each size category. It is possible to identify firms at different stages of dynamism (measured solely in terms of housebuilding output). Four categories, based on recent performance, can be identified. The relevant parameters of each category were as follows:

Annual variation in output (1979–82)

(1) Stable	< ± 15%	
	or ± 1 dwelling	whichever the greatest
(2) Expanding	> + 15%	
	or + 2 dwellings	whichever the greatest
(3) Contracting	> − 15%	
	or − 2 dwellings	whichever the greatest
(4) Fluctuating	> ± 15%	
	or ± 2 dwellings	whichever the greatest

Although increases in output of up to 15 per cent in a company with more than several hundred annual dwelling completions might warrant an expanding classification, no firms of this size showed growth of this magnitude – growth where evident amongst such companies was on a much larger scale.

In overall terms, it is evident that despite the continuing recession most housebuilders have been able to sustain output, and a surprisingly high proportion (25 per cent) have actually managed to significantly increase production (Table A1.3). Analysis by housebuilder size, however, reveals a more intricate pattern.

TABLE A1.3: Classification of housebuilders by size and stage of dynamism

	Stable	*Expanding*	*Contracting*	*Fluctuating*	*Total*
Very small	8	–	–	2	10
Small	3	1	2	3	9
Medium	5	1	–	1	7
Large	4	8	1	1	14
Total	20	10	3	7	40

Of the 'very small' housebuilders, 80 per cent were classified as having a stable annual output. This is not so surprising since any substantial increase or decrease would remove the firm from this size category. However, even the intentions of these firms were seldom other than to maintain a steady output: only one had expectations of expansion when market conditions were deemed favourable and only one had the clear intention of withdrawing from housebuilding entirely. For the rest, output was stable and these closely resemble what Craven (1970) aptly labelled the 'stagnant entrepreneurs'. The reasons for their willingness to persist or reluctance to deviate from this level of output are not entirely explained by financial factors.

The attitudes, motivations and fears of the one-man builder-developer are of prime significance. Their sense of independence is highly developed and, for a few of the sample, this was highlighted by a desire to free the business of the ties of persistent loan requirements. This imposes a severe impediment to growth.

For many more, there was a reluctance to expand because of the extra management and loss of direct control this would entail. Ambition was frequently stunted by pride and independence.

Added to this was the element of fear. This fear derived from personal experience or knowledge of colleagues who had attempted to expand but 'overstretched' themselves and verged upon bankruptcy. In the sample, three of the very small housebuilders had actually come very close to bankruptcy in the slumps of 1973/4 or 1977/8. Actual experiences or tales of such persist in many minds and go a long way in explaining the caution of many in the industry. The psychological impact of the 1973/4 slump will persist for many years to come.

Amongst the 'small' housebuilders, fewer firms might be classified as 'stable' but 33 per cent were labelled as 'fluctuating' and 22 per cent as 'contracting'; only one firm appeared to be 'expanding'. This pattern seems to reflect the nature of the recession for housebuilders in Central Berkshire: the very small housebuilders can continue to find and finance single- and double-dwelling plots and the larger developers can operate on the large-release sites but the small housebuilders tend to find problems in securing sites of operable size. None the less, the flexibility of the small housebuilder should not be underestimated. This size of firm is usually prepared to tackle a wide range of dwelling types and prices and, while completions might fluctuate, an analysis of annual turnover might show a more consistent pattern. Moreover, the small housebuilders often have other concerns to which resources may be switched.

The 'medium-sized' housebuilders generally have a more stable completion rate. One is actually expanding, though not at a rate which would be likely to necessitate an imminent reclassification. Certainly, none wished to enter the 'large' company league in the foreseeable future. But a tension does exist amongst some of these companies and, because traditional land banks are running out and because planning and general housebuilding management has become more sophisticated, a few may have difficulty in sustaining current levels of output. The traditional family firms are those most likely to suffer – the newer consortia appear more willing to progress via large site development and by general diversification of company interests.

In contrast, the 'large' companies gave an impression of confidence and competitiveness. Most were expanding but in part this is merely a reflection of selective migration. As demonstrated in the section on operational area (p. 60), the large companies have been attracted to Central Berkshire by its status as a growth area and by the availability of land. Obviously, successful firms will tend to be those who are most likely to take up such opportunities: to expand operational area is to increase output. However, another important factor is the battle at the top to become 'Britain's largest housebuilder'. This battle is geared around number of annual completions and not profit. There is considerable prestige in being recognized as a major housebuilder. To be regarded as a major housebuilder is perceived as an asset to other company interests.

Appendix 2 Planning applications for new dwellings in Central Berkshire 1974–81

As an area of high demand for residential development, a high proportion (41 per cent in the last quarter of 1981) of the planning applications submitted in Central Berkshire concern new dwellings.

It is clear from Table A2.1 that, when standardized by area, the number of new dwelling applications in Central Berkshire is more than twice that of England as a whole. The pressure upon land is high. However, when standardized by head of population, the number of new dwelling applications in the study area is slightly below that of England as a whole – but the latter standardization is a poorer indicator of current pressure because dense populations represent past development pressures.

TABLE A2.1: New dwelling application decisions issued in last quarter of 1981 in Central Berkshire and in England

	Central Berkshire	*England*
Applications	126	16,123
Applications/100 hectares	0.26	0.12
Applications/100 people	0.33	0.35
% approved	69.8	74.4

Between 1974 and 1981, the local authorities of Central Berkshire dealt with 5,471 new dwelling applications. However the sizes and success rates of these proposals varied throughout the area. It gives a useful insight of each of the districts to analyse these differences.

Within the four local planning authorities of Central Berkshire, development pressure as standardized by area is greatest in Reading and least in Newbury (see Table A2.2). But the sizes of the applications differ in each district.

Reading, despite its urban nature, has a surprisingly high number of fairly large applications. But this should not imply that Reading has many developable large sites. Certainly, some exist (the largest being the Tilehurst Potteries site of eighteen hectares) but many of the applications for six or more dwellings often refer to flat developments. Reading acknowledges that it has a shortage of housing and it is therefore to be expected that developers will exploit such a situation to press for high-density development. Hence the relatively low number of single-dwelling applications in the borough.

Newbury East, in contrast, has the highest proportion of single-dwelling applications. Since much of Newbury East has Area of Outstanding Natural Beauty (AONB) status and since most is classed as countryside, only infilling and rounding-off is permitted within structure plan policies. For much of the area, then, a single- or double-dwelling application is the only viable option. However, there are some large sites to the west of Reading and to the south, around Burghfield and Mortimer, there has been pressure for large-scale development. Nearly all of the large-site applications concern these areas.

Wokingham District Council is an area of sharp contrasts from the highly restrictive Green Belt in the north to the massive Lower Earley development in the west. Its application size profile is similar to that of Newbury East but, importantly, over 16 per cent of all its applications concern more than ten dwellings (and very often more).

Finally, Bracknell District Council has many fewer single-dwelling applications and, appropriate to the new town status of its centre, it has a very high proportion of large site applications. The very major sites in Bracknell and to the south in Sandhurst and Crowthorne stand in stark contrast to the virtually unpressurized Crown Lands south east of the new town.

Applicants do not submit proposals in a vacuum; through policy statements and pre-application negotiations planning officers have an influence upon the applications submitted. The discrepancy between applicants and officers' perceptions are perhaps best demonstrated by Table A2.4 which indicates the success rate of applications in each district. Before analysing Table A2.4 in detail it is first necessary to indicate the different likelihood of success of different types of application (Table A2.3). Detailed applications have higher success rates. The most contentious issue is most commonly established at the outline stage. Once outline

TABLE A2.2: Residential planning applications for Central Berkshire districts 1974–81

	Total applications	*Applications per 100 people per year*	*Applications per 1,000 people per year*	*Size by dwelling*		
				Single	*2–9*	*10+*
Reading BC	810	2.74	0.74	40.7	35.3	16.4
Newbury DC	1,182	0.88	3.87	61.9	25.6	12.4
Wokingham DC	2,587	1.81	3.20	59.5	24.2	16.2
Bracknell DC	892	1.02	1.52	50.1	26.2	23.7
Central Berks	5,471	1.39	1.80	55.7	26.5	16.7

permission is approved, it should be easier to agree upon the details of a proposed development and there should be greater rapport between applicant and planner encouraging even more pre-application negotiation over areas of disagreement.

TABLE A2.3: Success rate (%) of new dwelling applications, by size and type, in Central Berkshire (excluding Reading Borough Council)

	Size (by dwelling)			*Total*
	Single	*2–9*	*10+*	
All applications	56.7	55.2	56.4	56.3 (4,661)
Detailed applications	72.9	76.8	77.1	74.5 (2,229)
Outline applications	41.6	38.6	33.4	39.6 (2,432)

TABLE A2.4: Success rates of new dwelling applications in Central Berkshire 1974–81, by district and size

ALL APPLICATIONS					
	Size (by dwelling)			*All*	*Total number*
	Single	*2–9*	*10+*		
Reading BC	57.0	52.4	54.9	55.2	810
Newbury DC	62.4	64.0	65.3	63.2	1,182
Wokingham DC	52.8	49.4	51.4	51.8	2,587
Bracknell DC	60.8	59.0	60.2	60.2	892
Central Berks	56.7	54.6	56.2	56.1	5,471
OUTLINE ONLY					
Reading BC	45.6	37.9	43.5	43.2	359
Newbury DC	45.3	58.0	47.2	46.3	607
Wokingham DC	38.6	33.7	28.4	35.9	1,352
Bracknell DC	46.1	39.7	33.0	41.4	473
Central Berks					2,791

As can be seen from Table A2.4, there are quite marked differences in the success rates of applications in each of the districts. No specific explanation can be offered, but it is useful to identify some of the major factors which may have an influence.

First, the emphasis of planning policy in each district differs. Perceptions of these policies also vary. It might be hypothesized that where planning policies are more clearly defined there will be fewer speculative or unacceptable applications; where some confusion exists, the best test is to submit an application. Notably, in casual conversations, the officers at Bracknell and Newbury have volunteered that their policies are very clear and they do indeed show the highest success rates.

The planning policies of an area must have some effect upon the type of developer attracted. With its very large sites, Wokingham District Council has attracted at least five of the top ten housebuilders in the UK. Having been drawn to an area, experienced developers have an interest in exploring the limits of planning policy in the area. One way of doing this is to submit what are fully realized to be contentious planning applications. Sustained pressure, supported by appeals which go against the local planning authority's stance, may ultimately lead to more favourable decisions for developers. Such a confrontational process, however, will be accompanied by an apparently low application success rate.

While, on the input side, a developer's perceptions of and incentive to change policies will have an effect upon success rates, it is also to be expected that the stance of the decision makers will have an influence. The significance of the pre-application discussions has been noted. Through such discussion, development control officers will implicitly or explicitly present their interpretation of policy. The personality of the officer is likely to be important. Where policy is clear, even the most deferential officer should be able to provide an authoritative and unambiguous statement of an applicant's likely success. Conversely, where policy is changing or vague, even the most forceful officer will find difficulty in providing a definitive statement – rather than trying to discourage an applicant, (s)he will recommend submission.

Appendix 3 The local planning authorities

The planning process is not easily penetrated; often only the outcomes are recorded fully in written form. The reasoning behind decisions is often shrouded and effectively hidden from public view. A variety of information sources have been used to investigate the local planning system, which fall into five major categories: documentary sources, a survey of public committee meetings, a questionnaire survey of elected representatives, observation of private local planning authority meetings, and participant observation of development control in action in one district council.

Documentary sources

Four different documentary sources provide information on planning applications and decisions. (i) The minutes of public and private committee meetings record the decisions taken and resolutions made. These are useful for tracing the course of particular applications, but shed little light on the accompanying discussion.

(ii) The planning registers of each authority record the key aspects of all planning applications submitted: the application description, site address, applicant's name and address, the date of receipt and determination by the authority, the decision, and, where applicable, the outcome of any appeal. In Bracknell District Council the register also indicates the recommendation of the planning officers. A study of the differences between officer recommendations and committee decisions was undertaken using the Bracknell register, for the period 1974–81 (see Fleming and Short, 1984). In the text this survey is referred to as the Bracknell Survey.

(iii) Every application received by a local authority is allocated a case file within which is kept relevant documentation and correspondence, and ultimately a copy of the decision letter. Although not public documents, no problems were encountered regarding access to these files. They were used to provide further information on officer–committee divergences in the Bracknell Survey, and were an important data source for the case studies presented in Chapter 7.

(iv) The fourth documentary source was the written report to the committee for each application. The nature of these reports are explained more fully in the text, but the salient characteristics will be noted here. They are issued to councillors seven to ten days before committee, and accompany the agenda for the meeting. In three of the four districts (excluding Wokingham) considerable detail is presented for development control items. This includes the policy context of the application, the planning issues raised, the views of statutory and sometimes non-statutory consultees, the number of written objections and the principal points involved, the officer's view on the application, and the recommended decision, with conditions or refusal reasons as appropriate. Technical, policy and design issues are generally reported well but the political implications of development are rarely considered. The written reports were used principally as the source of the recommendations, although an amended or reversed recommendation was sometimes verbally given at the committee meeting. Wokingham District Council operated with a different system, private subcommittees and verbal reports, and the planning committee agendas contained reports (and recommendations) for only the major applications and the council's own proposals.

Committee meetings

To gain information about planning committee debate and dynamics, and the public interaction between councillors and officers, a series of public planning committee meetings were attended. Much of the public discussion of planning matters centres upon development control issues, and full notes were made of committee debate. This allowed analysis of the discussion times for different application types and between authorities, the topics discussed and the precise comparison of recommendations and decisions. The survey covered three meetings in each of the four districts between November 1982 and February 1983. The exception was Bracknell District Council, whose short development control agendas prompted coverage of an additional meeting to augment the original sample (for the recommendation-decision comparison only). Wokingham's private subcommittee structure prevented a full four-authority comparison of decisions and recommendations; this part of the analysis is thus based only on Reading, Bracknell and Newbury (the Three District Survey).

In addition to this basic three-cycle survey, a much larger record of committee meetings was made in two authorities: Reading and Newbury. This involved fourteen meetings in Reading between November 1982 and March 1984, comprising seven meetings before and after the May 1983 local elections when the Conservatives gained overall control of the council, and eighteen in Newbury.

Questionnaire survey

The above data sources left many questions unresolved, and to gain a clearer understanding of the actions of district planning councillors, a questionnaire survey was undertaken. Lists of planning councillors are easily obtained from the yearbooks maintained by every authority. These books indicate (amongst other things) the name, address, party (although not in all cases), ward, and committees/subcommittees of each member. The sample was drawn from these books: each planning chairperson was interviewed and a stratified (by district and party) random sample of the other planning councillors was obtained. A one-third sampling fraction was used; the nineteen councillors comprised ten Conservatives, three Labour, three Liberal and two SDP members. The questionnaire included sixty-three principal questions, covering fourteen aspects of the role, background, attitudes and actions of district planning councillors. To obtain a wide range of tabulated data from the survey, most questions employed a set of fixed responses. Many carried more open-ended supplementary questions, however, and verbatim replies were noted whenever possible. Respondents were also given two tables related to time spent on activities and pressure groups which they filled in and returned by post. The quotations by councillors used within the text, and also in the councillor profiles, are derived from questionnaire replies. The minimum length of interview was forty minutes, the maximum two and a half hours, with an average of approximately one and a half hours.

Private meetings

The survey of committee meetings examined the public face of planning, but many of the meetings held during the life cycle of a planning application are in private session. Building on the contacts established by councillor and officer interviews, access was gained to ten private meetings. Two types of forum were common to all the districts; committee site visits (two attended) and agenda briefings for the committee chairperson and vice chairperson (three attended). Three other committees were each unique to one of the Central Berkshire districts; the development control working party for large residential sites (Newbury – two attended), the planning committee party group meeting (Reading – one attended), and the private development control subcommittee (Wokingham – two attended). These occasions also presented opportunities to informally discuss with councillors and officers the actions taken at the meetings. The responses and impressions obtained from formal interviews and public meetings were therefore supplemented by observations of the planning process in action and casual questionning of the participants.

Participant observation

The fifth approach to understanding the local planning system involved part-time work in one district planning department by one member of the research team for a period of one month. This participant observation exercise permitted much greater understanding of the attitudes, actions and roles of the planning officers, and the relationships between them and various client groups: councillors, other local authorities, applicants and agents, and the public. During this period the progress of two important residential applications was observed, including the discussions with developers, consultations with county council departments and the involvement of councillors. Private and public committee discussion was subsequently noted, together with the developers' perspective upon the process. Other important insights gained from this source concerned pre-application discussions between officers and developers/agents/planning consultants, and the involvement of councillors at an early stage in the process.

Appendix 4 Parish council survey

The analysis of parish activities in Central Berkshire was conducted at two levels. First, there was a general survey of the forty-five parishes of Central Berkshire. A standard letter was mailed to each parish clerk and from the forty-three replies a picture of the structure, composition and committees of parish councils was compiled.

The second level consisted of an interview with twelve councillors from different parish councils. Of these twelve, nine were either current or past chairpersons of the council while the other three were chairpersons of the planning committee. The distribution by type of parish was as shown in Table A4.1. Our interviewed sample would thus seem to be skewed towards the experience of developed parishes.

TABLE A4.1 The parish survey in relation to the parish development typology

	% distribution	
	Surveyed parishes (n)	*All parishes*
Stable	8 (1)	28
Developed	50 (6)	28
Wavering	34 (4)	33
Developing	8 (1)	11
Total	100 (12)	100

Appendix 5 Residents' associations

Generating the sample

A variety of sources were used to identify contact points for residents' groups in Central Berkshire. The *Directory of Local Organizations*, available from local libraries, listed various local groups, societies and organizations including playgroups, sports clubs, social groups and residents' associations. The County Planning Department and each District Planning Department in Central Berkshire maintains a list of amenity societies and residents' groups who have asked to be sent copies of planning applications and policy documents. This was our major source of information. A standard letter was also sent to all local parish councils asking for assistance with the names and addresses of residents' groups. These three sources covered existing groups. In order to obtain information on groups operating in the recent past but now defunct the Civic Trust in London was contacted to obtain information on groups' operation in Central Berkshire in the recent past and a search was also made through copies of two local newspapers – the *Reading Chronicle* and the *Evening Post* – back to 1974 for mention of any residents' group activity. Extra groups were also identified in the course of interviewing residents' groups. A total of 149 groups were identified and contacted. There were twenty-four refusals and in twenty-five cases contact was made but a questionnaire was not completed because of difficulties in arranging suitable times for interview. In eight cases the groups were found to be general amenity or environmental societies. The interviewed sample of residents' groups thus consists of ninety-two residents' groups.

The questionnaire

The aim of the questionnaire was to elicit general information on residents' groups and particular information on the aims, organization and action of each group and their interaction with other agents involved in the production and planning of the built environment. The questionnaire was part formal in that a number of questions with precise answers were called for, and informal in that a number of prompts were used to elicit general discussion. No question seemed to cause a consistent difficulty in eliciting a response. The interviews ranged from forty-five minutes to two and a half hours in duration and were held with a central member, generally the secretary of the residents' group. Apart from written material (such as minutes, etc.) available on the group most information on the group which we obtained was mediated through only one individual. Ideally, a number of both central and peripheral members should have been interviewed for each group. However, this was beyond the resources of the project. The partial nature of the information should be borne in mind. Information was most easily obtained on general background questions such as data formation; the main problem area was in the attempts to measure the frequency and effectiveness of the group's actions. Few people could put an exact or even approximate figure to different types of action, and group spokespersons could rarely tell if their actions alone had achieved anything. We persevered with this question however since it seemed crucial to our understanding of residents' groups and their effectiveness.

The member profile

In order to supplement information gathered through the survey of residents' groups we did a further survey of twenty-one key individuals in separate residents' groups. The sample was not random. The main criterion was to question those individuals who had been forthcoming in the initial interview. The second criterion was to establish a rough parity between the member survey and the general survey in terms of type of group. Of the twenty-one members interviewed the distribution by type was as shown in Table A5.1.

The member survey consisted of general, open-ended questions allowing the respondents to develop their ideas. The technique used was the drop and collect method whereby respondents were given the question sheets which were then collected a week later. From the twenty-one replies, nineteen were usable in terms of empirical analysis. One of the 'refusals' is shown as a block, in Chapter 6.

TABLE A5.1 Breakdown of resident group surveys

	% distribution	
	Member survey	*General survey*
Stopper	45	43
Getter	25	37
Stopper-getter	30	20
Total	100	100

Bibliography

ALEXANDER, A. (1982), *The Politics of Local Government in the United Kingdom*, Longmans, Harlow.
BALL, M. (1983), *Housing Policy and Economic Power*, Methuen, London.
BARTLETT SUMMER SCHOOL (1979), *The Production of The Built Environment*, University College, University of London.
BARTLETT SUMMER SCHOOL (1980), *The Production of The Built Environment*, University College, London.
BARTLETT SUMMER SCHOOL (1981), *The Production of the Built Environment*, University College, University of London.
BATHER, N. J. (1973), 'Residential location: the spatial distribution of new growth and the role of the developer in a sub-regional housing market', unpublished PhD thesis, University of Reading.
BERKSHIRE COUNTY COUNCIL (1976), *Central Berkshire Structure Plan: Report of Survey*, County Planning Office.
BERKSHIRE COUNTY COUNCIL (1977). *Central Berkshire Structure Plan: Consultation Document*, County Planning Office.
BERKSHIRE COUNTY COUNCIL (1978), *Central Berkshire Structure Plan: Report of Public Participation*, County Planning Office.
BERKSHIRE COUNTY COUNCIL (1980), *Central Berkshire Structure Plan: Approved Written Statement*, County Planning Office.
BERKSHIRE COUNTY COUNCIL (1982), *Central Berkshire, Land for 8,000 Houses: The Choices*, County Planning Office.
BHASKAR, R. (1979), *The Possibility of Naturalism*, Harvester Press, Hassocks, Sussex.
BLONDEL, J. and HALL, R. (1967), 'Conflict, decision making and the perceptions of local councillors', *Political Studies*, 15, 338–43.
BOURDIEU, P. (1977), *Outline of a Theory of Practice*, Cambridge University Press.
BOURDIEU, P. (1980), *Le Sens Pratique*, Editions Minuit, Paris.
BRAY, C. (1981), 'New villages: case studies no. 1', *Working Paper no. 51*, Department of Town Planning, Oxford Polytechnic.

BREHENY, M. J. (1983), 'A practical view of planning theory', *Environment and Planning B*, 10, 101–15.

BREHENY, M., CHESHIRE, P. and LANGRIDGE, R. (1983), 'The anatomy of job creation? Industrial change in Britain's M4 corridor', *Built Environment*, 9, 61–71.

BUTTERFIELD, H. (1950), *The Whig Interpretation of History*, G. Bell & Sons, New York.

CARDOSO, A. and SHORT, J. R. (1983), 'Forms of housing production: initial formulations', *Environment and Planning A*, 15, 917–28.

CASTELLS, M. (1983), *The City and The Grassroots*, Edward Arnold, London.

CORINA, L. (1974), 'Elected representatives in a party system: a typology', *Policy and Politics*, 3(1), 69–87.

CRAVEN, E. (1970), 'Conflict in the land development process: the role of the private residential developer', unpublished PhD thesis, University of Kent at Canterbury.

CROSSMAN, R. (1975), *The Diaries of a Cabinet Minister*, Jonathan Cape, London.

CULLINGWORTH, J. B. (1975), *Environmental Planning, Vol. 1, Reconstruction and Land Use Planning 1939–1947*, HMSO, London.

CULLINGWORTH, J. B. (1981), *Environmental Planning, Vol. 4, Land Values, Compensation and Betterment*, HMSO, London.

DAMESICK, P. (1982), 'Strategic choice and uncertainty: regional planning in southeast England', in Hudson, R. and Lewis, J. R. (eds), *Regional Planning in Europe*, Pion, London.

DAVIS, J. and HEALEY, P. (1983), 'Wokingham: the implementation of strategic planning policy in a growth area in the South-East', *Working Paper no. 74*, Department of Town Planning, Oxford Polytechnic.

DEARLOVE, J. (1979), *The Reorganisation of British Local Government*, Cambridge University Press.

DOE. (1980a), Circular 9/80: *Land for Private Housebuilding*.

DOE. (1980b), Circular 22/80: *Development Control – Policy and Practice*.

DREWETT, R. (1973), 'The developers: decision processes', in Hall, P. *et al.* (eds), *The Containment of Urban England, Vol. 2*, Allen & Unwin, London.

ELCOCK, H. (1982), *Local Government: Politicians, Professionals and the Public in Local Authorities*, Methuen, London.

FLEMING, S. (1984), 'Housebuilders in An Area of Growth: Negotiating the Built Environment of Central Berkshire', *Geographical Papers*, no. 84, University of Reading.

FLEMING, S. and SHORT, J. R. (1984), 'Committee rules OK. An examination of planning committee action on officer recommendation', *Environment and Planning A*, 16, 956–73.

FORREST, R. (1983), 'The meaning of home ownership', *Environment and Planning D: Society and Space*, vol. 1, 205–16.

FRIEDEN, B. J. (1978), *The Environmental Protection Hustle*, MIT Press, Cambridge, Mass.

FRIEDMAN, M. and FRIEDMAN, R. (1980), *Free To Choose*, Secker & Warburg, London.
GALBRAITH, J.K. (1958), *The Affluent Society*, Hamish Hamilton, London.
GIDDENS, A. (1981), *A Contemporary Critique of Historical Materialism*, Macmillan, London.
GORDON, I. (1979), 'The recruitment of local politicians: an integrated approach with some preliminary findings from a study of Labour councillors', *Policy and Politics*, 7, 1–37.
GREEN, D.G. (1981), *Power and Party in an English City: An Account of Single Party Rule*, Allen & Unwin, London.
GREGORY, D. (1981), 'Human agency and human geography', *Transactions of the Institute of British Geographers*, new series, 6, 1–18.
GREGORY, D. (1982), *Regional Transformation and Industrial Revolution*, Macmillan, London.
GYFORD, J. (1976), *Local Politics in Britain*, Croom Helm, Beckenham, Kent.
HALL, S. and JACQUES, M. (eds) (1983), *The Politics of Thatcherism*, Lawrence & Wishart, London.
HARVEY, D. (1978), 'Urbanization under capitalism: a framework for analysis', *International Journal of Urban and Regional Research*, 2, 101–31.
HARVEY, D. (1982), *The Limits to Capital*, Basil Blackwell, Oxford.
HAYEK, F.A. (1982), *Law, Liberty and Legislation*, Routledge & Kegan Paul, London.
HEALEY, P. and UNDERWOOD, J. (1977), 'Professional ideals and planning practice', *Progress in Planning*, vol. 9, part 2.
HENRY, D. (1982), 'Planning by agreement in Berkshire District', *Working Paper no. 69*, Department of Town Planning, Oxford Polytechnic.
HILL, M.J. (1972), *The Sociology of Public Administration*, Weidenfeld & Nicolson, London.
JARMAN, O. (1979), 'Building production in a capitalist economy: the response of a sample of building companies to changing market conditions', unpublished MSocSc. thesis, University of Birmingham.
JOHNSTON, R.J. (1979), *Political, Electoral and Spatial Systems*, Oxford University Press.
JOINT LAND REQUIREMENTS COMMITTEE (1983), *Is There Sufficient Housing Land for the 1980s? Paper II: How Many Houses Have We Planned For: Is There a Problem?*, Housing Research Foundation, London.
KIRBY, A.M. (1982), *The Politics of Location*, Methuen, London.
KNOX, P.L. and CULLEN, J.D. (1981), 'Planners as urban managers: an exploration of the attitudes and self-image of senior British planners', *Environment and Planning A*, 13, 885–98.
LEOPOLD, E. and LEONARD, S. (1983), 'Reorganised labour', *Building*, July, 25–6.
LOWE, P. and GOYDER, J. (1983), *Environmental Groups in Politics*, Allen & Unwin, London.

MACKIE, J. (1982), *Goodbye Rural Berkshire*, Binfield & Warfield Parish Councils.

MASSEY, D. and CATALANO, A. (1978), *Capital and Land*, Edward Arnold, London.

MAUD, SIR JOHN (Chairman) (1967), *Report of the Committee on The Management of Local Government*, HMSO, London.

MORGAN, C.L. (1982), 'A sutdy of the Lower Earley Community', unpublished BA dissertation, University of Reading.

MORRIS, D.S. and NEWTON, K. (1970), 'Profile of a local political elite: businessmen as community decision-makers in Birmingham 1838–1966', *New Atlantis*, 1, 111–23.

MCNAMARA, P. (1982), personal communication.

NEWTON, K. (1974), 'Role orientations and their sources among elected representatives in English local politics', *The Journal of Politics*, 36(3), 615–36.

NEWTON, K. (1976), *Second City Politics*, Oxford University Press.

NICHOLS, D.C., TURNER, D.M., KIRBY-SMITH, R. and CULLEN, J.D. (1981), *Private Housing Development Process: A Case Study*, Department of the Environment, London.

PAHL, R. (1975), *Whose City?*, Penguin, Harmondsworth.

PINCH, P. (1984), *Parishes and Planning in Berkshire*, mimeo, Dept. of Geography, University of Reading.

PLANNING ADVISORY GROUP REPORT (1965), *The Future of Development Plans*, HMSO, London.

PLANNING INSPECTORATE (1982), *The Chief Planning Inspector's Report for 1982*, HMSO, London.

PLENDER, J. (1982), *That's the Way the Money Goes: The Financial Institutions and the Nation's Savings*, André Deutsch, London.

RAY, M. (1983a), 'Controlling development and planning ahead', *Town and Country Planning*, 52(4), 96–8.

RAY, M. (1983b), personal communication.

REDCLIFFE-MAUD, LORD (Chairman) (1969), *Royal Commission on Local Government in England*, 3 vols, Cmnd 4040, HMSO, London.

RED RAG (1983), 'Planning in the Golden Triangle', part 1.

REES, A.M. and SMITH, T. (1964), *Town Councillors: A Study of Barking*, Acton Society Trust, London.

ROBINSON COMMITTEE (1977), *Committee of Inquiry into the System of Remuneration of Members of Local Authorities*, Cmnd 7010, HMSO, London.

ROBSON, B.T. (1982), 'The Bodley barricade: social space and social conflict', in Cox, K. R. and Johnston, R. J. (eds), *Conflict, Politics and the Urban Scene*, Longmans, Harlow.

RYDIN, Y. (1983), 'Housebuilders as an interest group: the issue of residential land availability', *Geography Discussion Papers*, new series, 6, London School of Economics.

SANDBACH, F. (1980), *Environment, Ideology and Policy*, Basil Blackwell, Oxford.

SAUNDERS, P. (1979), *Urban Politics*, Hutchinson, London.

SAUNDERS, P. (1980), 'Local government and the state', *New Society*, 13 March 1980, 550–1.

SHORT, J.R. (1982a), *An Introduction To Political Geography*, Routledge & Kegan Paul, London.

SHORT, J.R. (1982b), *Housing in Britain: The Postwar Experience*, Methuen, London.

SHORT, J.R. (1984), *The Urban Arena*, Macmillan, London.

SHORT, J.R., FLEMING, S. and WITT, S. (1984), 'New pressures on London's Green Belt', *The Geographical Magazine*, 56, 90–2.

SKEFFINGTON, ARTHUR (Chairman) (1969), *Planning and People*, Report of the Committee on Public Participation in Planning, HMSO, London.

SMYTH, H. (1982), 'Land banking, land availability and planning for private house building', *Working Paper no. 23*, School for Advanced Urban Studies, University of Bristol.

STODDART, R. (1983), 'Structure plans in Berkshire – theory and practice', in Cross, D.T. and Bristow, M.R. (eds), *English Structure Planning*, Pion, London.

THOMPSON, I. (1981), 'The Lower Earley development brief', unpublished MPhil dissertation, University of Reading.

THRIFT, N. (1983), 'On the determination of social action in space and time', *Environment and Planning D*, 1, 23–58.

UNDERWOOD, J. (1980), 'Town planners in search of a role', *School for Advanced Urban Studies Occasional Paper no. 6*, School for Advanced Urban Studies, University of Bristol.

WATES, N. (1976), *The Battle For Tolmers Square*, Routledge & Kegan Paul, London.

WITT, S.J.G. and FLEMING, S.C. (1984), 'Planning councillors in an area of growth: little power but all the blame?', *Geographical Paper no. 85*, University of Reading.

Index

NOTES 1 All references are to Central Berkshire, unless otherwise indicated. 2 Sub-entries are in alphabetical order, except where chronological order is significant.

accountability of planning committee, 150, 161
active residents' associations, 206–7
activist councillors, 126, 177, 179
administrative staff in building firms, 67–8
advertising: houses, 77, 78–81; land, 49–50; *see also* media
advice, *see* information
age: of district councillors, 119–20; of parish councillors, 176; of residents' associations' members, 225
aggressive housebuilder, 110, 111–12
aims, *see* motivations
airport, 30; *see also* transport
Aldermaston, 87, 164, 168, 174, 190, 197
Alexander, A., 8
amelioration factor in facilitation, 155
ameliorative planning gain, 267–8
amenity groups, *see* residents' associations
AONB, *see* Area of Outstanding Natural Beauty
appeals, 94–7, 153–4
applicants' pressure on councillors, 138–9, 140
applications, planning *see* planning authorities
architectural agencies and consultants, 48, 107–8, 114
Area of Outstanding Natural Beauty, 163, 191, 286
area planning assistants, 98
Ascot Society, 210, 212, 229, 232
Ashampstead, 163, 164, 168, 174, 190
Association of Municipal Corporations, 8
autonomy of builders, 62

Ball, M., 37, 278
banking, land, 53–6
banks, 83, 86–7
bargaining between authorities and developers, 23–4
Barlow Report on Distribution of Industrial Population (1940), 14
Baron, T., 23, 89, 91, 94, 113
Barratt, L., 113
bartering, planning, 268
Bartlett Summer School, 37
Bather, N. J., 45, 278

Beansheaf Farm, 75
Berkshire, *see* preliminary note
Berkshire Association of Local Councils, 193, 243
betterment, 16
Bhaskar, R., 3
Binfield, 34, 164, 168, 174, 190, 193, 221, 234; Village Protection Society, 232, 240–7
blocking in public committee meeting, 152, 153–4
Blondel, J., 121
boroughs, 7–10; *see also* Bracknell; Newbury; Reading; Wokingham
boundaries, 7–11, 25; *see also under* maps
Bourdieu, P., 3
Bracknell district, 10, 11, 25; campaigns against development, 240–7; development near, 71, 75; development pressure, 24, 26, 27; parish councils, 164, 168, 170–1, 174–5, 183–4, 190, 192, 194; planning applications, 228, 286–7; planning authorities, councillors, 102, 116–19, 123, 134; planning authorities, system in action, 141–4, 149, 151–4, 156, 159; residents' associations, 219, 221, 232, 236
Bradfield, 164, 168, 170, 174, 190, 197
Bray, C., 74, 80
Breheny, M. J., 5, 24
British Design Standards Authority, 78
builder-developers, *see* housebuilders
building societies, 72, 83
Burghfield, 27, 164, 165, 168, 174, 190, 197, 221, 234
Burghfield Common, campaign against development, 247–57
Butler, T., 270
Butterfield, H., 6
Byron, Lord, 7
California Residents' Association, 222, 224, 232
'call-in' procedure, 94
capital: customers', 83–5; housebuilders', 37, 38, 85–8; *see also* expenditure; finances; investment
Cardoso, A., 37
case studies and examples, 240–65; Binfield and Warfield, 240–7; Burghfield Common, 247–57; councillors, 123–4, 128, 130, 131, 149, 180, 182, 196; housebuilders, 45, 53, 64, 73, 88; interest registers, 123–4; large building site, 76–8; Newhurst Nursery, 258–65; parish councils, 180, 182, 196; planning, 100, 123–4, *see also* councillors *above*; planning gains, 269, 270–1; planning pressure, 136, 140; residents' associations, 211, 213–15, 217–19, 235, 237, 239
Castells, M., 1, 204
cautious housebuilder, 109–10
Caversham, 164, 168, 174, 190; Residents' Association, 34, 222–3, 232
Caversham Park, 164, 168, 174, 190; Village Association, 212, 218, 229–30, 232
Central Land Board, 15–16
central government: and housebuilders, 93–7; – local government relations, 8–9, 12–13, 15, 272–5
chairpersons: briefing, 146–7; in parish councils, 170, 176–80
Chestnuts Residents' Association, 222, 224
Chief Planning Officer, 92, 98, 243
choice, customers' lack of, 84
civic societies, *see* residents' associations
Civic Trust, 230, 295
clearing banks and capital provision, 86–7
clerk of parish council, 169–70

Coley Park Residents' Association, 212
collective response to planning, 159–61
communications, *see* transport
community, 16, 18; response to planning, 271–2; *see also* residents' associations
Community Land Act (1975), 16
commuting, 30; *see also* transport
companies, housebuilding, 41–4; *see also* large; medium-sized; small housebuilders
compensatory planning gain, 268
computer industries, 27, 30
concurrence in public committee meeting, 151–3
conditional contract and land purchase, 54–5
Confederation of British Industries, 90
Conservatives: and economic development, 13; and land use planning, 15, 16, 20, 22–3; and local government re-organization, 8–10; and local planning authorities, 116–17, 119, 121, 126, 134, 253, 259–60, 266; and parish councils, 172–3, 175, 178; *see also* politics
consortia, housebuilding, 41, 43–4; *see also* large housebuilders
constituency speeches, public statements as, 158
consultants, 48, 107–8, 114
Consultation Document, The, 189–90
consumers: capital provision, 83–5; responsiveness to, 84–5
control over details, 159–60
Corina, L., 122
costs, *see* capital; expenditure; finances; investment
'council filler' role of councillors, 128, 131–2, 180
Council for Protection of Rural England, 210, 230, 235, 243–4
councillors, district, 115, 116–39; characteristics of, 119–22; housebuilders and, 101–4; motivations, 122–7; party politics and, 116–19, 133–4; pressures on, 132–40; roles of, 127–32; *see also* development control; planning authorities
councillors, parish, 175, 182; characteristics of, 176–80; and residents' associations, 231, 236; *see also* parish counties, 7–9, 18, 97–8
CPRE *see* Council for Protection of Rural England
Craven, E., 45, 47, 283
Crossman, R., 8, 9
Crowthorne, 164, 168, 174, 190, 221, 286
Cullen, J. D., 115
Cullingworth, J. B., 14, 15, 18
customers, *see* consumers
cycles: land bank, 53, 56; land search, 49–53

Davis, J., 107
decision-taking in planning committee, 150–9
Dee Road Residents' Association, 212, 215, 229–30
defence industries, 27, 30
deferral in public committee meeting, 152, 156–7
deflection of blame as collective response, 159–60
defunct residents' associations, 206, 218, 220
delegation, 141–2
Department of Environment: appeals to, 94–6; circulars, 20–3, 91–4, 244, 250, 255, 263; and Green Belt, 35, 113; and housebuilders, 88, 89, 91, 93–7, 113; *see also* 'Heseltown'; Secretary of State
design of houses, 78, 84, 188
'developed' parishes, 163, 164, 172, 197, 220, 294
developers, *see* housebuilders
'developing' parishes, 164, 165, 172, 197, 220, 294
development control system, 139,

141–59, 276; delegation agreements, 141–2; early approaches, 142–3; written reports, 143–4; personal investigation, 144–5; recommendation, 145; pre-committee meetings, 145–50; public committee meeting, 150–9, 291–2; *see also* councillors, district; planning authorities
development plans, 15, 17, 47–8; *see also* structure plan
development pressure, 24–35
development proposals and parish councils, 181, 183–202
discussion in public committee meetings, 157–9
distribution: of residents' associations, 219–22; *see also* maps
district councils, case studies of tensions, *see* Binfield; Burghfield, Newhurst district planning, *see* development control; planning authorities
DOE, *see* Department of Environment
dormant residents' associations, 206–7, 218, 220
Dower Report on National Parks (1945), 14
Drewett, R., 45, 46
dual representation on parish and district councils, 198–9
dynamics of housebuilders, 281–2

Earley, 164, 168, 174, 190, 221
Easthampstead, 164, 168, 170, 174, 190
educational qualifications of district councillors, 120
Elcock, H., 122
elected councillors, *see* councillors
Eliot, T. S., 37
employment: in housebuilding, 66–8, 279; in industries, 27, 29
enhancement of environment, *see* 'getters'
enterprise zones, 13
environment, concern with, *see* residents' associations
Erleigh Road Residents' Association, 212, 229
Essex Design Guide, 78
estate agents: house sales and, 81–3, 88, 114; land sales and, 49, 50–1, 114
examples *see* case studies
expenditure: builders', 60; on infrastructure, 74, 267–8; local government, 12–13; public, 30, 266–7; *see also* capital; finances; investment

facilitation, 152, 154–6
family advice centres, 18
family businesses in housebuilding, 41, 42–3; *see also* medium-sized: small
Federation of Master Builders, 89, 113
finances of residents' associations, 229; *see also* capital; expenditure; investment
Finchampstead, 164, 168, 174, 190, 221
Fleming, S.: on committee response, 151, 290; on councillors, 119, 132; on facilitation, 154, 156; on housebuilders, 41; on overturned recommendations, 101
'flexi-home', 84
FMB *see* Federation of Master Builders
Fords Farm Residents' Association, 216, 229–30
Frieden, B. J., 209
Friedman, M. and R., 266
functionalist fallacy, 3–4

gain, *see* hard gain; planning gain; soft gain
Galbraith, J. K., 267
gearing ratio of builders' costs, 86
'gentleman's agreement', 81
get-involved councillors, 126–7, 177, 179
'getters' residents' associations,

212–15, 218, 220, 225, 228, 231, 238–9, 297
Giddens, A., 3
goals, *see* motivations
Gordon, I., 119
government, *see* central government; local government
Goyder, H., 226
Green, D. G., 150
Green Belt, 35, 40, 113, 163, 191, 200
Gregory, D., 2, 4
growth strategies, 31, 32–4
Gyford, J., 122

Hall, R., 121
Hall, S., 266
hard gain for councillors, 122–5, 177–8
Harvey, D., 265
Hayek, F. A., 266
HBF, *see* Housebuilders' Federation
Healey, P., 107, 115
Henry, D., 76, 268
Heseltine, M., 22–3, 35, 89, 94–5, 240, 241, 242
'Heseltown' (proposal for extra 8000 houses), 35, 273–4; campaigns against, 240–3, 246–52; district councils and, 129; land for, 54, 97; parish councils and, 124, 189–93, 201; residents' associations and, 234–6
high density housing, 70
high growth strategy, 31, 32–4
high technology industries, 27, 29–30
higher authority, public statements aimed at, 159–60
Hill, M., 122
Home Office community development projects, 18
housebuilders, 19–23, 37–114, 276–84; stages and agents, 38–40; general context, 40–1; company context, 41–4; land assembly, 37, 38, 44–60; production of housing, 39, 60–71; marketing and selling new housing, 37–8, 39, 78–83, 113–14; provision of customers' capital, 83–5; capital provision for, 37, 38, 85–8; external contacts, 30–1, 88–92; planning system and, 13–14, 30, 39, 92–112, 271; emerging trends and, 112–14; sample, 276–84; organizations, 89–92, *see also* Housebuilders' Federation; case studies of tensions, *see* Binfield, Burghfield, Newhurst
Housebuilders' Federation, 79, 89, 90–2, 276; and DOE, 94, 95; Heseltine on, 22–3; and land release, 56; and planning, 88, 97, 113; and timber frame dwellings, 72
housebuilding, *see* case studies; housebuilders; parish councils; planning; residents' associations
houses: extra 8000, *see* 'Heseltown'; prices, 69–70, 209; *see also* housebuilders
Housing Act (1974), 23–4
Housing Research Foundation, 85
human factor in facilitation, 155
Humber, R., 91
Hurst Village Society, 34, 209–10, 231, 232, 233, 236

ideology of planning, 13–15, 114
images of house and area, 78–9
incentives to buy new houses, 79–81
incorporated public groups, 106
Independents in parish councils, 172–3, 175
individual entrepreneurs, 41, 42; *see also* small housebuilders
individual response to planning, 159–61
individuals' pressure on councillors, 135
industry, *see* economic development
information: about land, 49, 50, 51; and parish councils, 183–4;

and residents' associations, 183–4
infrastructure, 38, 101, 267–8
insurance industry, 26
interest, vested, 122–4, 177–8
international companies, 26
investment, 26, 27, 38, 101; *see also* capital; expenditure; finances
Issues Report, The, 189–90

Jacques, M., 266
Johnston, R. J., 10
Joint Land Requirement Committee, 55

Katesgrove, 164, 168, 174, 190: Residents' Association, 216, 229, 232, 234, 236
Kidson, P. L., 29
Kirby, A. M., 204
Knox, P. L., 115

labour, *see* employment
Labour Party: and economic development, 13; and land use planning, 16, 20; and local government reorganization, 8–10; and local planning authorities, 117–20, 126, 134, 259–60, 264, 266; and parish councils, 173, 175; *see also* politics
land: assembly, 37, 38, 44–60; banking, 53–6; nationalization, 14–15; prices, 50–1, 55, 59, 69, 81, 82, 85; release, 13, 20–2; search, 45–9, 97
Land Commission Act (1967), 16
Land for 8000 Houses: the Choices, see 'Heseltown'
land use planning system, 13–24; for reconstruction, 14–15; reconstruction of, 16–24; *see also* planning authorities
large housebuilders, 280–2, 284; capital provision, 85–6, 87; case study, 88; customers' capital, 83; HBF and, 89–90; housing production, 61–9, 70–1, 73, 76–8; housing sales, 80, 82; land assembly, 46, 48, 49–51, 52, 54–6, 57; planning system, 95–6, 101, 105, 107–8
large sites, 74–8; *see also* Lower Earley
legislation, 15–16, 23–4; *see also* Local Government Acts; Town and Country Acts
leisure complex, 269
Leonard, S., 66
Leopold, E., 66
Liberal Party: and planning councillors, 116, 118, 119–20, 126; in parish councils, 173, 175; *see also* politics
life cycle of residents' associations, 206–8, 218–19, 220
limited growth strategy, 31, 32–4
loans, *see* capital
Local Authority Act (1981), 12
local government, 7–13; expenditure, 12–13; reorganization (1974), 9–10, 11, 18, 102, 181
Local Government Acts: (1888 and 1894), 10, 165; (1958), 8; (1972), 7, 10, 12, 18, 123, 141, 166, 185
Local Government Planning and Land Act (1980), 12, 16, 256
local inquiry appeal strategy, 96
local plan, 17, 233
local planning authorities, *see* development control; planning authorities
local planning response, 31–5
Location of Industry Act (1945), 15
Lowe, P., 226
Lower Earley, 74–9, 92, 103, 137, 186, 232, 233

M4, 24, 25, 29, 30, 40; *see also* roads; transport
McDonald, G., 29
Mackie, J., 244, 245
Mallaby Committee, 8
'management committees', 106
maps: boundaries, 11, 25, 75, 164,

168, 174, 190; communications, 25; local government, before and after reorganization, 11; location of Central Berkshire, 3; parishes, 164, 168, 174, 190, 221; Reading, 223; regional context, 3; residential sites, major, 75; residents' groups, 221, 223, 224; Wokingham, 224
marketing and selling of new houses, 37–8, 39, 78–83, 113–14
Marlborough Avenue Residents' Group, 222, 223
maternalist councillors, 126, 177, 178–9
Maud, Sir John and Report on Local Government (1967), 8, 119, 149, 176
Meadow Residents' Association, 34
media: and economic development, 24, 29; on housing estates, 77; and residents' associations, 237–8; on timber frame construction, 72, 74; *see also* advertising
medium-sized housebuilders, 41, 43, 280, 282–4; case studies, 64, 73; capital provision, 86; customers' capital, 83; housing production, 61, 63, 67–9, 70; land assembly, 46, 47, 50–1, 52, 56; and planning system, 96, 103, 108
members, *see* councillors
members of parliament, lobbying, 92, 95
membership of residents' associations, 222–8
merchant banks and capital provision, 87
metropolitan areas, 9–10
micro-chip industries, 27
Ministry of Housing and Local Government, 17, 18
mixed housing, 71, 78
modification in public committee meeting, 152, 156
Morris, D. S., 123
motivations: of councillors, 122–7, 178–80; of members of residents' associations, 219, 224–5

naïve housebuilders, 110–11
National Federation of Building Trades Employees, 89–90, 91
National Federation of Community Organizations, 230
National Housebuilding Council, 68, 72, 85, 90, 91, 276
National Parks, Conservation and Access to Countryside Act (1949), 15
national planning response, 31
nationalization of land, 14–15
negotiating housebuilders, 110, 111
neighbourhood groups, *see* residents' associations
New Ash Green, 80, 83
New Homes Marketing Board, 79, 85, 92
New Towns Act (1946), 15
Newbury East district, 10, 11, 25; development pressure, 24; housing sites near, 75; parish councils, 164, 165, 167–8, 170–2, 174, 183–4, 190, 194, 196; planning applications, 228, 285–8; planning authorities, councillors, 116, 118–19, 123; planning authorities, system in action, 142–3, 146, 148–9, 152–3, 156, 161
Newhurst Nursery, 258–65
Newton, K., 123, 127, 129
NFBTE, *see* National Federation of Building Trades Employees
NHBC, *see* National Housebuilding Council
NHMB, *see* New Homes Marketing Board
Nichols, D. C., 86

occupations: of district councillors, 120; of parish councillors, 177, 178; of residents' associations' members, 225, 226–7

office development, 26, 28–9, 121, 126, 134, 160
officers, 115; and chairperson's briefing, 146–7; delegation to, 141–2; and housebuilders, 98–101; informal consultation with, 144; and parish councils, 193–6, 198–202; pressure on councillors, 132–3; recommendations, 145; and site visits, 148
operational areas of housing production, 60–6
opportunist strategy of land search, 48–9
option agreement and land purchase, 54–5, 272
Options Document, *see* 'Heseltown'
Orwell, G., 115
output of housebuilders, 282–4
outright purchase of land, 55
Owlsmoor, 75, 232
owner-occupation, 203; *see also* housebuilders

Padworth, 164, 168, 174, 190, 197
Pangbourne, 164, 168, 174, 190, 196, 197, 221
para-careerist councillors, 125, 177
parish and parish councils, 10, 12, 163–202, 294; campaigns against development, 24–65, *see also* Binfield, Burghfield, Newhurst; factor in facilitation, 155; and housebuilders, 93, 104–5; organization, 165–71; and planning, 175–80, 181, 183–202; politics and, 171–81, 195, 198–200; pressure on councillors, 135–7; and residents' associations, 220–2, *see also* campaigns *above*
Parliamentary Liaison Team, 92
participation, 18–24, 181
parties, *see* politics
party group pre-meeting, 149–50
paternalist councillors, 126, 177, 178–9
permission, planning, 15, 58–60, 98
personal investigation and planning, 144–5
physical: environment, concern with, 208, *see also* residents' associations; infrastructure, 38, 101, 267–8
'place agent' role of councillors, 128, 129–31, 135, 142, 161–2, 179–80
Planning Advisory Group, 17, 18
planning agencies, 107–8, 114
planning authorities/applications, 10, 18, 23–4, 30, 115–62, 285–9; data sources, 290–3; and housebuilders, 13–14, 30, 39, 92–112, 271; and parish councils, 175–80, 181, 183–202; and residents' associations, 228, 231, 233–6; *see also* councillors; development control; land use; politics
planning bartering, 268
planning consultants, 48
planning gain, 16, 23–4, 74, 267–72; collective response, 159, 160–1
planning permission, 15, 58–60, 98
planning profits, 58
planning response to development pressure, 31–5
planning status of purchased land, 58–60
Plender, J., 26
point of sale, 81–3
'policy administrator' role of councillors, 127–9, 142, 161, 179
political speeches, public statements as, 160
politicization of housebuilding, 112–13
politics: and economic development, 13; and land use planning, 15–16, 20, 22–3; and local government reorganization, 8–10; and local planning authorities, 116–21, 126, 133–4, 253, 259–60, 264,

266; and parish councils, 171–81, 195–6, 198–200; and residents' associations, 231–3; *see also* Conservatives: Labour; Liberal; Social Democratic
population growth, 40
precepts, *see* rates
pre-committee meeting, 145–50
pressure: on councillors, 132–40; development, 24–35
prices: house, 69–70, 209; land, 50–1, 55, 59, 69, 81, 82, 85
private: gain and public planning, 16; landowners, 49, 50–1, 52; planning, 145–50; property, 15, 16; sector growth and public sector retrenchment, 265–7
production of housing, 39, 60–71; operational areas, 60–2; Central Berkshire as operating area, 63–6; labour component, 66–8, 279; dwellings constructed, 68–71; timber frame, 71–4, 113–14; large sites, 74–8
professional politician councillors, 127, 177
professionalization of housebuilding, 112
profits: house sale, 70; planning, 58; turnover, 57, 279–80
protection of environment, *see* 'stoppers'
public: accountability, 150, 161; collective response, 159–60; expenditure, 30, 266–7; inquiries, 233; limited companies, 41, 42, 44, 114; participation, 18–19, 181; planning and private gain, 16; planning committee meeting, 150–9, 290–1; sector retrenchment and private sector growth, 265–7; statements in planning committee, 157–9
publicity, *see* advertising; media
purchase of land, *see* land search
Purley, 164, 168, 174, 190, 196, 197
quality of housing, 84–5; *see also* design

railways, 30, 40, 269; *see also* transport
rates, 170–1, 181, 266
Ray, M., 156
Reading Borough (district), 10, 11, 25; development pressure, 24, 26, 27, 28–9; house prices, 69; housing sites near, *see* Lower Earley; office boom, 26, 28–9, 121, 126, 134, 160; parish councils, 164, 168, 174, 190; planning applications, 228, 286–7; planning authorities, councillors, 116–19, 121, 123, 126, 134; planning authorities, system in action, 142–3, 146, 148–50, 152–3, 156–62; residents' associations, 219, 221, 222–3, 229–30, 232; transport links, 40
reasons for formation of residents' groups, 208–18
recommendation, planning, 145
Redcliffe-Maud, Lord and Royal Commission on Local Government (1969), 8, 9, 10, 18, 163, 166, 172
Rees, A. M., 122, 178
regional: context of Central Berkshire, 3; councils, proposed, 9; limitation of land search, 45–6
Reith Report on New Towns (1946), 14
release, land, 13, 20–2
Remenham, 163, 164, 168, 174, 190, 221
reorganization, *see under* local government
Report on Survey, The, 189–90
residents' associations/community groups, 203–39, 295–7; group formation, 205–22; internal resources, 222–30; external contacts, 30–1, 106–7, 137–8, 228, 230–8; consequences of, 238–9; case studies of tensions,

see Binfield; Burghfield; Newhurst
resistance to development, 159–60; *see also* residents' associations
resource environment, 208
restricted growth strategy, 31, 32–4
roads, 268, 270–1; *see also* M4; transport
Robinson Committee Report (1977), 119, 149
Robson, B. T., 208
roles: councillors', 127–32, 177–8; in residents' associations, 219
Royal Commission, *see* Redcliffe-Maud
rural mythology, 79, 209
Ruscombe, 164, 168, 174, 190, 221
Rydin, Y., 89

St Nicholas Hurst, 164, 168, 174, 190, 221
sales, *see* marketing; prices
Sandbach, F., 203
Sandford, *see* Woodley
Sandhurst, 75, 164, 168, 174, 190, 221, 286
saturation strategy of land search, 46–8
Saunders, P., 123, 275
Scott Report on Land Utilization in Rural Areas (1942), 14
search, land, 45–9, 97
Secretary of State, 76, 94–5, 273; *see also* Department of Environment; Heseltine
self-financing capital provision, 85–6
selling houses, *see* marketing
service industries, 29
Shinfield, 164, 168, 174, 190, 221, 235
Short, J. R., 4; on committee response, 151, 290; on facilitation, 154, 156; on Green Belt, 40, 113; on modes of housing production, 37; on overturned recommendations, 101; on residents' associations, 203, 205; on small builders, 70; on suburbanization from London, 40
site visits, 144, 147–9
size of building sites, 46; *see also* large sites
size of building firms, 46; *see also* large housebuilders; medium-sized; small
Skeffington, Arthur and Report (1969), 19
small housebuilders, 41, 42, 280, 282–3; capital provision, 85–6, 87; case studies, 45, 53; customers' capital, 83; housing production, 61, 63, 66, 67–9, 70; land assembly, 46, 50–1, 52, 56, 57; and planning system, 95, 96, 103, 105, 108, 110–11
Smith, T., 122, 178
Smyth, H., 53, 57
social councillors, 125–6, 177
Social Democratic Party, 116, 118–19, 175; *see also* politics
social environment, 208
social infrastructure, 38, 267
soft gain for councillors, 125–7, 177–80
Sonning, 164, 168, 174, 190, 221
South East Study, The, 31
speculative house production, *see* housebuilders
Spencers Wood, 233, Residents' Association, 234–6
'stable' parishes, 163–5, 172, 197, 220, 294
stages and agents in residential development, 38–40
stalling in public committee meetings, 152, 156–7
status/class, *see* occupation
status-tician councillors, 125, 177
Stoddart, R., 134, 273
'stoppers' residents' associations, 208–12, 220, 225, 227–31, 238, 297
'stoppers-getters' residents' associations, 215–18, 219–20, 228, 230, 297

Strategic Plan for South-East, The, 31, 76
Stratfield Mortimer, 164, 168, 174, 190, 196, 197, 221
Streatley, 164, 168, 174, 190, 197
structure plan, 17, 19, 66, 273–4; boundaries, 25; and DOE, 94; and HBF, 91; and residents' associations, 233; response, initial, 34; strategies, 31–5; *see also* development plans; 'Heseltown'
success rates in planning applications, 109–12
Sulham, 164, 168, 174, 184, 190
Symons, A., 240

Tee, H., 269
tensions, emerging, 35–6, 265–75
Thatcher, M., 242, 266
Theale, 164, 168, 174, 190, 197, 221, 234
Thompson, I., 77
Thrift, N., 4
Tilehurst, 164, 168, 170, 190, 197, 221
timber frame construction, 71–4, 113–14
time commitment of councillors, 149
'top-up' sites, 56
Town and Country Planning Acts: (1947), 15–16, 17, 23, 92; (1968), 17–19, 23; (1971), 19, 23
transport and communications, 24, 25–6, 29–30, 40, 269; costs for builders, 60; planning gain, 268–9, 270–1
turnover profits, 57, 279–80
Twyford, 34, 164, 168, 174, 190, 221

'underpricing' housing, 82
Underwood, J., 115
unemployment, 30
unions, 90
Uthwatt Report on Compensation and Betterment (1942), 14, 15–16

vested interest, 122–4, 177–8
VHBSG, *see* Volume House-Builders, etc.
village envelope concept, 187
Volume House-Builders Study Group, 89, 91, 94, 113
voluntarism, dangers of, 3
voluntary organizations, *see* residents' associations

ward pressure on councillors, 134–9
Warfield, 164, 165, 168, 174, 179, 190, 193, 221: Amenity Society, 232, 240–7, 263; Newhurst Nursery development, 258–65
Wargrave, 164, 168, 174, 190, 221
Waterloo Residents' Association, 212
Wates, N., 205
'wavering' parishes, 164, 165, 172, 197, 200, 220, 294
White Papers, 9, 16
windfall, 16
Winkfield, 164, 168, 174, 190, 221
Winnersh, 164, 168, 174, 190, 221, 234
Witt, S. J. G., 119, 132
Wokingham district, 10, 11, 25; development pressure, 24, 26; housing sites near, 75, 137, *see also* Lower Earley; parish councils, 164, 168, 171–2, 174, 183–4, 190, 194; planning applications, 228, 286–7; planning authorities, councillors, 92, 102–3, 116, 119–20, 123, 127, 137; planning authorities, system in action, 140, 142–3, 149, 151–2, 158, 160; residents' associations, 34, 137, 219, 221, 222, 224, 232
women: in council, 120; in industry, 29; in residents' associations, 225
Woodley, 164, 168, 170, 172, 174, 190, 221; Airfield, 137
Woosehill, 75, 232, 233
written: report in planning, 143–4; representation appeal strategy, 96

For Product Safety Concerns and Information please contact our EU
representative GPSR@taylorandfrancis.com
Taylor & Francis Verlag GmbH, Kaufingerstraße 24, 80331 München, Germany

www.ingramcontent.com/pod-product-compliance
Ingram Content Group UK Ltd.
Pitfield, Milton Keynes, MK11 3LW, UK
UKHW022217190926
475702UK00003B/400

9780367772048